WHEN TO USE THIS STAR MAP:

Early March:	2 a.m.
Late March:	1 a.m.
Early April:	Midnight
Late April:	11 p.m.
Early May:	10 p.m.
Late May:	Dusk

This star chart is most accurate if used within an hour or so of the times listed and is plotted for observers located between 30° and 50° north latitude. All times are standard time; if daylight-saving time is in effect, add one hour.

To use this chart, hold it in front of you and rotate it so that the yellow label corresponding to the direction you are facing is positioned at the bottom, right-side up. The stars in the sky should match those depicted on the chart. The center of the chart is the zenith, the point in the sky directly overhead.

The circled numbers on the chart refer to the pages where objects in that region of the sky are described in this book. The numbers highlighted in red indicate the objects best seen at the times and dates listed above.

Facing North
Facing NW
Facing West
Facing SW
Facing South
CASSIOPEIA
PERSEUS
CEPHEUS
CAMELOPARDALIS
Capella
AURIGA
Polaris
URSA MINOR
Little Dipper
DRACO
LYNX
Castor
Pollux
GEMINI
Big Dipper
Mizar & Alcor
URSA MAJOR
LEO MINOR
CANCER
CANIS MINOR
Procyon
Zenith
CANES VENATICI
BOÖTES
COMA BERENICES
Sickle
LEO
Regulus
Denebola
Arcturus
SERPENS (CAPUT)
EQUATOR
SEXTANS
Alphard
HYDRA
VIRGO
Spica
ECLIPTIC
CORVUS
CRATER
CENTAURUS
© 2006 Sky & Telescope
Star magnitudes
−1 0 1 2 3 4

Binocular Highlights

99 CELESTIAL SIGHTS FOR BINOCULAR USERS

Binocular Highlights

99 CELESTIAL SIGHTS FOR BINOCULAR USERS

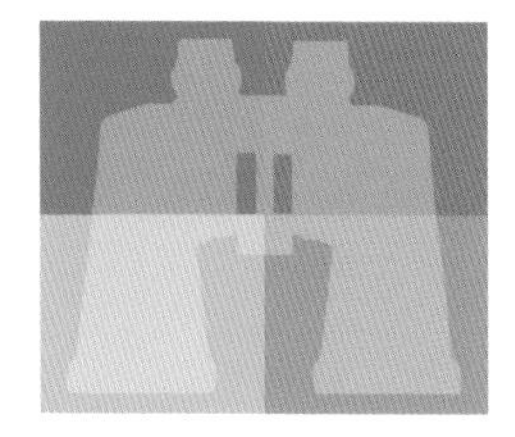

GARY SERONIK

NEW TRACK MEDIA LLC
Cambridge, Massachusetts
2006

Sky Publishing
49 Bay State Road
Cambridge, MA 02138-1200, USA
SkyTonight.com

Library of Congress Cataloging-in-Publication Data

Seronik, Gary.
Binocular highlights : 99 celestial sights for binocular users / Gary Seronik. — 1st ed.
p. cm.
ISBN-13: 978-1-931559-43-0
ISBN-10: 1-931559-43-0
1. Constellations — Observers' manuals. 2. Constellations — Atlases. I. Title.

QB63.S465 2006
523--dc22

2006021705

All charts by Gary Seronik, Gregg Dinderman, and Casey Reed. Printed in China.

TABLE OF CONTENTS

Akira Fujii

INTRODUCTION

It's July. The coals in the barbecue are starting to cool and the aroma of dinner still lingers in the warm twilight air. Glancing up, you notice a solitary star high overhead, and later another, and another. Soon, there are more points of light shining in the darkening sky than you can count. Perhaps you're noticing them for the very first time, or maybe the stars are as familiar to you as the birds and trees in your own backyard. It doesn't matter. For anyone susceptible to the charms of the night sky, twilight is a magic time, ripe with anticipation. If the sight of a starlit sky excites your curiosity, then this book is for you.

The night sky is full of wonders; some subtle, some grand. And you don't need a telescope to appreciate them. Within these pages you will find descriptions and finder charts for a variety of interesting objects that can be seen by anyone with binoculars. This collection is neither a "best of" (though many of the sky's finest deep-sky objects are included) nor a complete listing of everything within the grasp of ordinary binoculars. It is, however, a representative selection of star clusters, galaxies, nebulae, and double stars — all given the binocular highlight treatment.

But what is a *binocular highlight?* At its most prosaic level, Binocular Highlight is the name of a monthly column that appears in *Sky & Telescope* magazine. It's the place in the magazine where we introduce readers to wonders of the night sky that can be viewed with average binoculars, usually under less-than-perfect viewing conditions. The objects we profile each month also provide a jumping-off point to explore the universe from your backyard, and to learn a little about the craft of observing. Learning to really see is one reward for the effort and patience expended through many nights of poking around the constellations. But there is more to be enjoyed than just the satisfaction of finding some of the sky's many treasures.

When the view is fine and our imaginations engaged, we are

transported far from our daily cares and concerns, into the true vastness and awful indifference of the universe. It is humbling to gaze, for example, at a globular star cluster that is twice as ancient as the planet we inhabit. We are thrust directly to the limits of comprehension when we behold a nebula so vast that we have no earthly means of fully grasping its enormity. Our sense of time and space is subverted when we look at a distant galaxy whose light has traveled for millennia just to reach our eyes this particular night. And yet, as French art historian and writer André Malraux noted, "The greatest mystery is not that we have been flung at random between the profusion of matter and of the stars, but that within this prison we can draw from ourselves images powerful enough to deny our nothingness."

Take this book with you, along with your binoculars, and go out into fading twilight to begin your exploration. A universe full of wonder awaits.

Gary Seronik
Associate Editor
Sky & Telescope
June 2006

Sky Publishing: *Craig Michael Utter*

CHOOSING BINOCULARS

Just about everyone associates stargazing with telescopes. But even the most experienced backyard astronomer owns binoculars. Why? One reason is that when it comes to having a quick look at the sky, it doesn't get any easier than grabbing your binos — it's as close to instant observing as you can get. The other attraction is that ordinary binoculars take in a lot more sky than most telescopes. For big-picture views, binoculars can't be beat. And there are numerous objects so big you actually need binoculars to see them at their best. For beginners, there's the added appeal of low cost — decent binoculars are much less expensive than a good starter telescope. Also, unlike telescopes, which yield images that are upside-down or mirror-reversed, the view in binoculars matches the orientation of the real world. This means that the transition from unaided view to a magnified one is relatively uncomplicated.

Chances are you already own binoculars. Even if they've spent the last decade buried in the back of a closet and may not have the best optics in the world, you will still see more with them than with your eyes alone. That said, some binos are definitely better suited to stargazing than others.

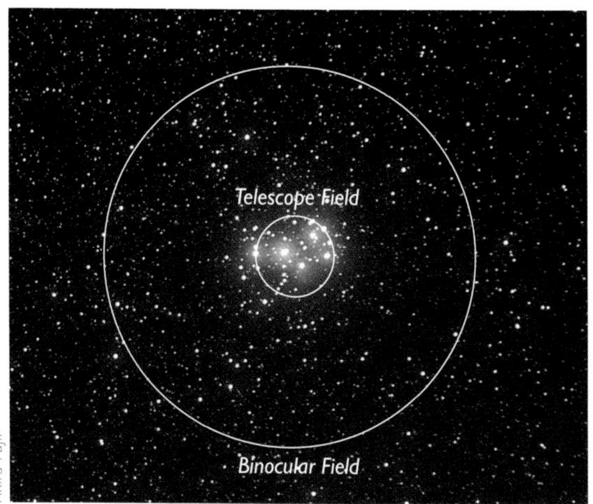

Akira Fujii

One advantage binoculars have over telescopes is their ability to show large swaths of sky. In this simulated view, you can see how the Pleiades Star Cluster will appear much more dramatic in 10 × 50 binoculars than in a typical telescope used at low power.

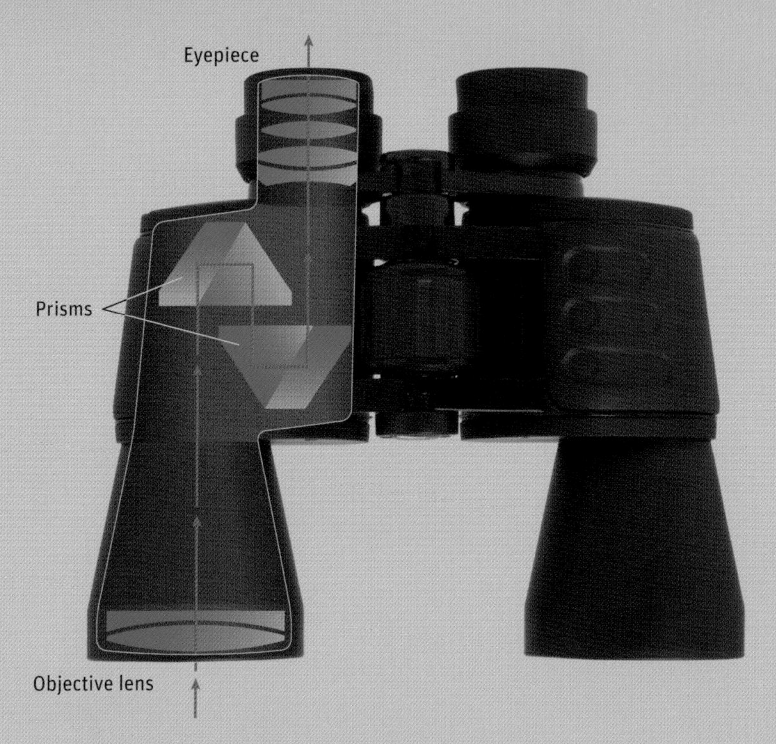

Binoculars consist of two identical sets of optics, each including an objective lens, internal prism assembly, and eyepiece. The objective lenses gather the light and the prisms direct this light to eyepieces, which magnify the image.

Sky Publishing: *Gregg Dinderman*

Decoding the Numbers

For the most part, binoculars are characterized by just two numbers: their magnification and the diameter of their two front lenses (called *objective lenses*). Fortunately, just about all binoculars have this information printed near the eyepieces. Typically, you'll see two numbers like 7 × 50, or 8 × 40, or something similar. The first is the *magnification factor* (how much closer objects appear than with the unaided eye), and the second is the diameter of the objective lenses, expressed in millimeters. For example, 10 × 50 binoculars will magnify by 10× (objects will appear 10 times closer) and have objective lenses that are 50 mm (about 2 inches) in diameter.

Most binoculars will also be marked with a second set of numbers that describe the width of the field of view. For example, you may see something like, "367 ft/1,000 yds." This means that if you were looking at a large building 1,000 yards away, you'd see 367 feet of its height and 367 feet of its width. But, when it comes to astronomy, where distances are usually in billions or trillions of

NUMBERS

Understanding the sets of numbers printed on the binoculars helps you assess their suitability for stargazing.

Sky Publishing: *Craig Michael Utter*

miles, this information has to be converted to a different set of measurements to be meaningful.

Astronomers measure sky distances in degrees (°). The distance from the horizon to the point directly overhead (the *zenith*) is 90°. The width of your fist on your outstretched arm is roughly 10°. The Moon is about ½° across. Luckily, converting a "ft/1,000 yds" number to degrees is merely a matter of taking the first number and dividing it by 52.4. So, the binoculars in this example will show 7° (367/52.4). If your binoculars use metric units — "112 meters at 1,000 meters" — then divide the first of these numbers by 16 instead.

Making Choices

So what combination of magnification and lens diameter works best for astronomy? The short answer is 10 × 50s.

And now, the long answer. In general, the bigger the objective lenses, the more light they collect — and collecting light is a big part of the game. Consequently, 50-mm binoculars are usually a better choice than 35-mm models. So if bigger is better, why not splurge for 70-mm or even 100-mm binoculars? Because large binoculars are heavy and tiring to use and usually have restricted fields of view. Years of stargazing have taught me (and many others) that 50-mm binos hit the sweet spot between capability and manageable size.

When it comes to the ideal magnification, the situation is a little more complex. In general, low-power binoculars show the most sky area — and more sky means your targets will be easier to find. However, most of the things you'll want to look at appear more conspicuous and detailed with higher magnification. So why not get 15× or 20× binoculars? Because as the magnification increases, the field of view shrinks until you reach a point where aiming the binoculars at a particular spot in the sky

WEIGHING THE NUMBERS

You can use the magnification and aperture numbers together to gauge the relative performance of different binoculars — simply multiply one number by the other. So, 10 × 50 binoculars would get a performance rating of 500, while 8 × 40s would only rate 320. In this case, the 10 × 50s should provide better views. Roy Bishop first proposed this method of evaluating binocular performance in the Royal Astronomical Society of Canada's *Observer's Handbook*, and my experiences closely agree with his ratings. The important point is that magnification and objective-lens size are interrelated — it isn't simply a matter of picking the biggest binoculars or the ones that magnify the most.

becomes a serious challenge. Also, as magnification goes up, the more difficult it becomes to hold the binoculars steady. All things considered, 10× seems to be the optimal magnification for binocular astronomy.

Other Features

A quick perusal of your local camera store's binocular section or advertisements on the Internet will bring you face to face with an overwhelming array of features. Fortunately, many of these can be safely ignored, but here are some things that are nice to have.

Tripod socket. Usually hidden under a plastic cap at the front of the binoculars is a socket threaded to mate with a right-angle binocular adapter that attaches to standard camera tripod. This makes it a breeze to mount your binos on a tripod or other support for steadier views.

Center-focus adjustment. There are two kinds of focusing schemes for binoculars — center focus (the most common) and individual eyepiece focusing. With center focus, you turn a dial located between the eyepieces to adjust the focus of both halves of the binocular simultaneously. This design is quick and easy to use. Although individual-focus eyepieces are mechanically simpler and generally more robust, focusing one eyepiece at a time can be tedious. Go with center focus.

On the other hand, here are a few things that don't matter (or at least usually aren't worth paying extra for).

Roof prism versus porro prism binoculars. Quality binoculars can be made in either design; one isn't inherently better than the other, though roof-prism models tend to be more expensive. Also, roof-prism binoculars that lack "phase coated optics" generally produce relatively dim, low-contrast views and are best avoided.

BK7 versus BaK4 prism. These terms refer to two different types of optical glass used in a binocular's internal prisms. While BaK4 prisms potentially offer better performance, the improvement is usually slight.

Big binoculars. Binoculars that feature objective lenses 70 mm and larger can provide amazing views — no question about it. That said, I have to confess that I'm not a fan. I've owned several pair over the years, but they always end gathering dust on a shelf. I find that their relatively small field of view, and the added hassle of setting up the required heavy-duty tripod and mount, completely offset the benefits. If I'm going to haul out that much equipment and get a field of only 3° or less, I'm going to opt for a telescope instead.

Attaching your binoculars to a camera tripod is easy if you use a right-angle tripod adapter, as shown here. Many binoculars have a ¼-20 threaded socket (usually concealed under a screw-on plastic cap) to accommodate such a device.

Sky Publishing: *Craig Michael Utter (2)*

Sky Publishing: Dennis di Cicco

Although big binoculars can deliver impressive views, their size and weight demand a sturdy mount and a heavy-duty tripod.

Binocular Optics: the Good, the Bad, and the Ugly

Stargazing places the greatest demands on binocular optics — there's nothing like a field of stars to make shortcomings alarmingly obvious. Binoculars that seem fine for bird watching or other daytime activities may not be stellar performers at night. So how do you go about choosing good optics? The best way is to try before you buy. If you perform two simple checks, you can safely avoid the most serious problems.

Test for sharpness. Center a star (or, if testing in the daytime, a glint of sunlight off a distant electrical pole insulator) and bring the binocular to a sharp focus. Next, slowly move the star (or sun glint) to the edge of the field of view. Does the star stay sharp and point-like, or does it fuzz out? Most binoculars will not produce sharp star images at the very edge of the field of view, but good ones will retain their sharpness over most of the field. Poor binoculars will show sharp images only at the very center of the field of view — and sometimes, not even there!

Testing for optical alignment. To prevent eyestrain (which can lead to headaches), both halves of a binocular need to be parallel and optically aligned. Mount the binocular on a tripod, or set it on a ladder or some other stationary platform. Aim the binocular at a *distant* building (at least several blocks away — the farther the better), and focus. Now, look in the right barrel only and note where objects are positioned relative to the edge of the field of view, along the horizontal (left-to-right) direction. Compare the view in the left barrel. Is everything in the same place relative to the edge of the field of view? Next, note the position of objects along the vertical (up-down) axis. Are they the same? Any binocular that shows a significant difference in image position between the left and right barrels should be rejected.

Of course there are many other factors that distinguish good binoculars from better ones, but as long as the optics are aligned *and* the image is sharp for most of the field of view, they will at least be usable.

The Allure of Cheap Binoculars

I once purchased a pair of perfectly serviceable 10 × 50 binoculars for $30 at an electronics retailer. These binoculars showed that if you choose carefully, you can get good optics for relatively little money. So what do you get if you spend ten times as much? In terms of the actual view, not as much as you might expect. Yes, more expensive binoculars have better optics that will deliver more light to your eyes and sharper images, but the difference is not night and day.

What the extra money does buy is mechanical quality. Expensive binoculars can withstand the inevitable bumps and knocks of everyday use without trouble, and have focusing mechanisms that are sure and precise.

Holding Steady: Binocular Mounts

Being able to hold your binoculars steady while viewing is just as important as the quality of the binoculars themselves. A shaky

COLLIMATION PROBLEMS

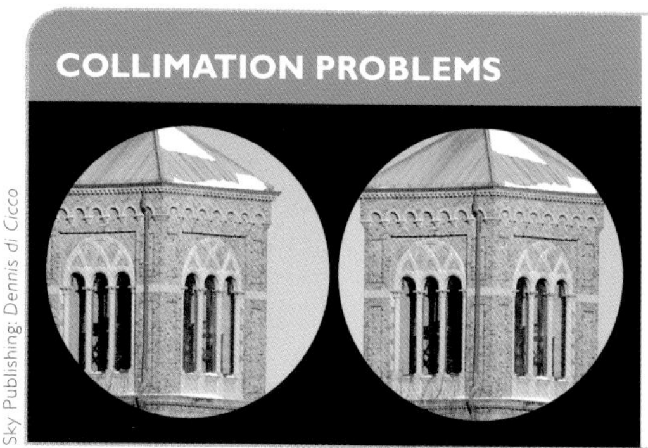

Sky Publishing: *Dennis di Cicco*

For a quick check of binocular alignment, compare the views produced by each binocular half and look for any image displacement. In this pair of images, you can see that the building appears shifted horizontally, indicating a serious problem.

view prevents you from seeing as much as the optics can deliver. Two factors play the biggest part in determining how steady the views will be: binocular weight and magnification. Heavy binoculars make your muscles work harder and shake more. And the higher the magnification, the more apparent the jiggling becomes. Indeed, even ordinary 10 × 50 binoculars benefit from some kind of support. Here are some suggestions.

A chair. Any kind of chair is better than none at all. Not only will you be more comfortable, but you'll be able to hold your binoculars steadier from a seated position. A reclining chaise lounge is best of all.

A camera tripod. This is only a partial solution. While a tripod does steady the view, for objects high in the sky things quickly get uncomfortable as you crane your neck to get under the binoculars.

A specialized binocular mount. These devices usually provide steady and comfortable views, but at additional cost and by sacrificing portability.

Sky Publishing: *Craig Michael Utter*

Binocular enthusiast Alan Adler suggested this elegant and inexpensive means of steadying binoculars — attach a piece of wood (a 6-inch length of half-round, in this case) to the top of a camera monopod. The binoculars are simply held against the crosspiece.

A camera monopod. This is my personal favorite. Used from a seated position it is possible to get steady and comfortable views. Adding a block of wood to the top and resting the binoculars against that (as shown at left) provides even better performance.

Image-Stabilized Binoculars

In my opinion, the best of all worlds is embodied in image-stabilized binoculars (ISBs). These ingenious devices incorporate extra optical elements and electronics to detect and compensate for image motion and the dreaded jiggles. The beauty of ISBs is that they provide won-

Sky Publishing: *Craig Michael Utter*

These image-stabilized units by Canon may be the ideal astronomy binoculars. The 10 × 42s (left) have superb optics and near-ideal specifications, while the smaller 10 × 30s combine good performance with a budget price.

derfully steady views without *any* extra gear. Essentially they preserve the true essence of binocular astronomy — the ability to go outside at a moments notice for a night of observing with a minimum of equipment and fuss.

Over the years, I have owned and tested many different ISBs. Of the current crop, my favorite is the 10 × 42 made by Canon. These combine superb optics with image stabilization and, in my view, are the ultimate tool for binocular stargazing. However, these wonder-binos aren't cheap — expect to pay more than $1,000.

For the budget-minded stargazer, Canon also makes a nice pair of 10 × 30 ISBs. These often retail for less than $300 and also feature excellent optics. Although their aperture is a little small, in field-testing I found they show as much as ordinary 7 × 50s. The Canons are small and lightweight enough to travel well, yet deliver performance quite beyond what you might expect for binos of this size. In fact, my 10 × 30s get used more often than any other binocular I own.

Final Thoughts

Although this chapter has been all about choosing binoculars, enjoying the night sky is *not* all about the equipment. Binoculars are simply a tool like any other. Keep in mind that any binocular is better than none. The main thing is to get outside and view the sky, to seek out its treasures, and soak up the starlight. When you're doing that, I guarantee that you'll be thinking about the beauty of the starry heavens, not the binoculars you're holding in your hands.

WINTER

Akira Fujii

1

DECEMBER • JANUARY • FEBRUARY

PLANETARY NEBULA
GLOBULAR CLUSTER
DIFFUSE NEBULA
OPEN CLUSTER
VARIABLE STAR
GALAXY

ABOUT THE CHARTS:

Each of the star maps in this chapter has been rendered at one of three different scales: the wide-field charts to magnitude 7.5, the medium-scale charts to magnitude 8.0, and the close-up charts to magnitude 8.5. Regardless, the darkened circular area always represents the field of view for typical 10 x 50 binoculars.

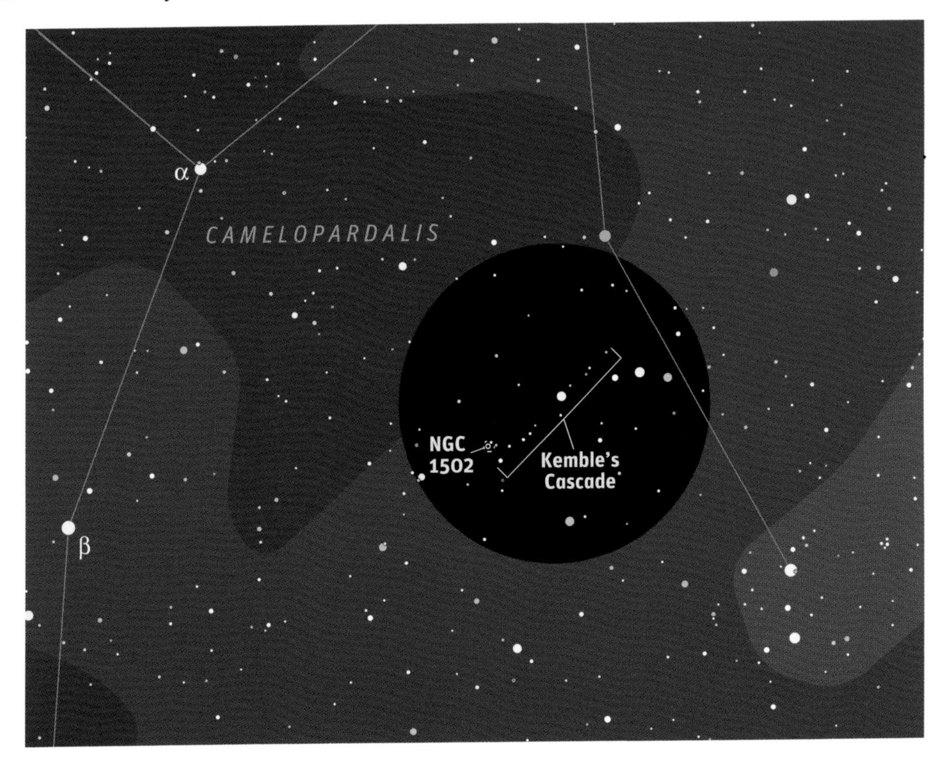

Kemble's Cascade

The eye and the brain are remarkably good at creating order from chaos. The constellations are one example, and so are the numerous asterisms (small star patterns within the constellations) that observers "discover" from time to time. However, the distinctiveness of a given pattern, like beauty, often lies in the eye of the beholder. One exception to this rule is the remarkable string of stars in Camelopardalis known as Kemble's Cascade.

I had my first encounter with this asterism in 1996 while examining a photograph of Comet Hyakutake that I had taken the previous night. Not far from the comet was an improbably long and straight line of a dozen or so stars of roughly the same brightness. Indeed, it looked so artificial that at first I dismissed it as a scratch on the negative!

This string of stars gets its name from the December 1980 Deep-Sky Wonders column in *Sky & Telescope* magazine, in which Walter Scott Houston recounted a letter he had received from Canadian amateur Lucian J. Kemble. In his letter Kemble described "a beautiful cascade of faint stars" in the constellation Camelopardalis.

Kemble's Cascade is visible even in light-polluted skies — I can make it out quite easily in 10 × 30 image-stabilized binoculars from my suburban backyard. Of course, darker skies improve the view considerably, since most of the stars in the group are between 8th and 9th magnitude. Under good conditions the cascade stands out better than the chart above suggests. As a bonus, Kemble's Cascade terminates near the open cluster NGC 1502. The cluster consists of a solitary 7th-magnitude sun surrounded by a tight fuzz of fainter stars and is a pretty easy find in the 10 × 30s.

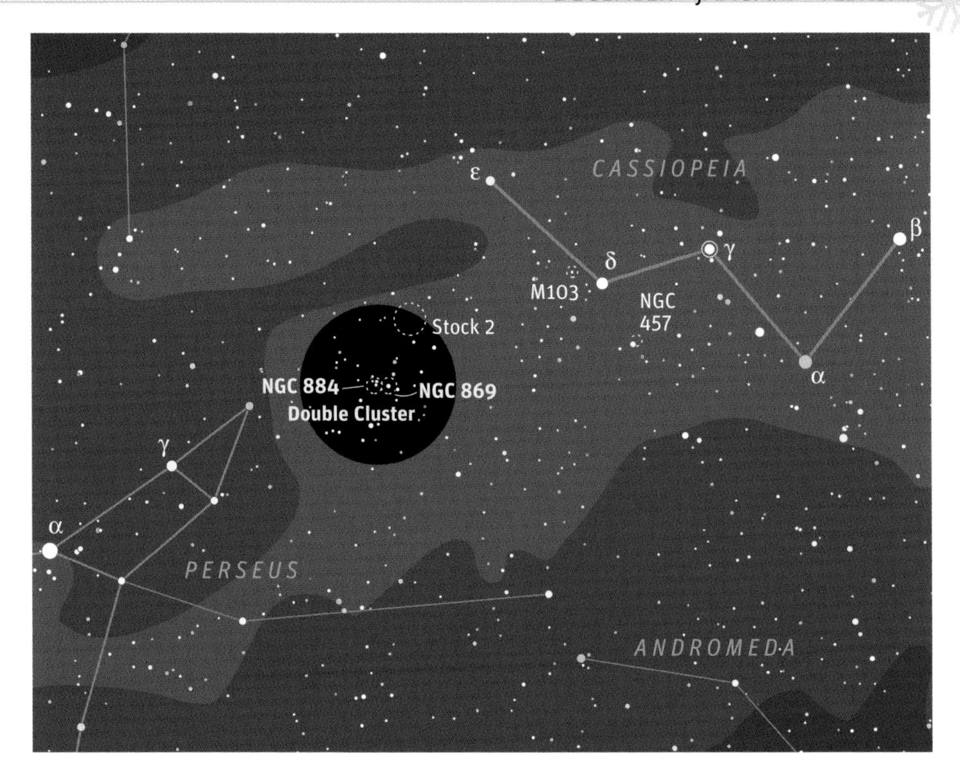

The Glorious Double Cluster

Without doubt, one of the most spectacular binocular sights in the sky is the famed Double Cluster in Perseus. For Northern Hemisphere observers, there are perhaps fewer than a dozen objects that are as impressive under typical light-polluted suburban skies. And when observed from dark-sky locations, these stellar mounds will convince even the most doubting observer that binocular astronomy has an appeal all its own.

The Double Cluster (also known as NGC 884 and 869) lies in the rich stretch of Milky Way between Perseus and Cassiopeia — appropriate when you keep in mind that open star clusters form near the plane of our Milky Way galaxy's disk. These two are about 7,600 light-years from us and only 13 million years old.

After soaking up the richness of the field of view, look carefully at the individual clusters and see if you can begin to notice differences between them. Is one more sparse than the other? Does one have more bright stars? Are they the same general shape? Trying to answer questions like these will help improve your observational skills and make it possible for you to see more difficult targets with greater ease.

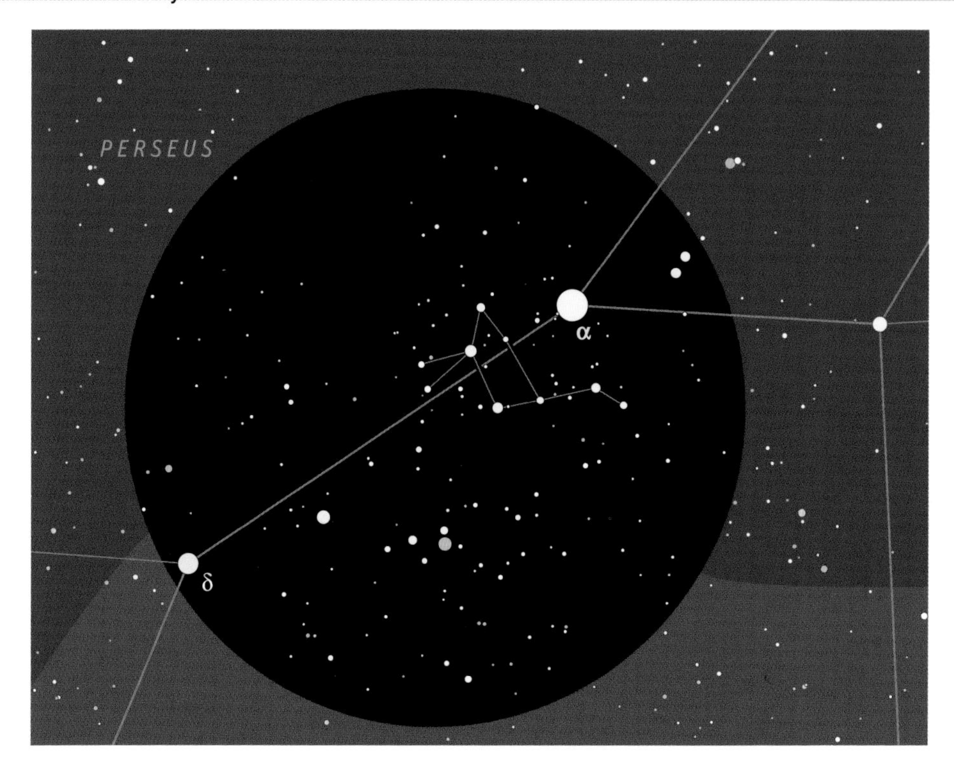

The Alpha Persei Association

Gatherings of stars come in all sizes. Some are tiny clumpings that require dark skies and large telescopes to be seen at all; others appear as impressive arrays of individual stars in binoculars or even without optical aid. The collection of stars that includes Alpha (α) Persei is one of the latter. It makes for a fine binocular sight on winter evenings, particularly for observers who have to contend with bright city skies. And this is no random, line-of-sight collection either. By measuring the distances to these stars and the direction of their motion, astronomers can tell that this is a stellar association — a swarm of young stars similar to an open cluster but larger and not gravitationally bound together.

The Alpha Persei Association features two dozen or so members brighter than 7th magnitude in an elongated, 3° patch of sky lying between Alpha and Delta (δ) Persei. It is a striking binocular field. Maybe it's because I'm a Canadian, but I can't help but trace the outline of a long-necked Canada goose whenever I view this cluster. Our eyes and brains come with a deep, inborn tendency to perceive, or imagine, patterns in the random noise of chaos. Perhaps you will see some form of your own in these stars.

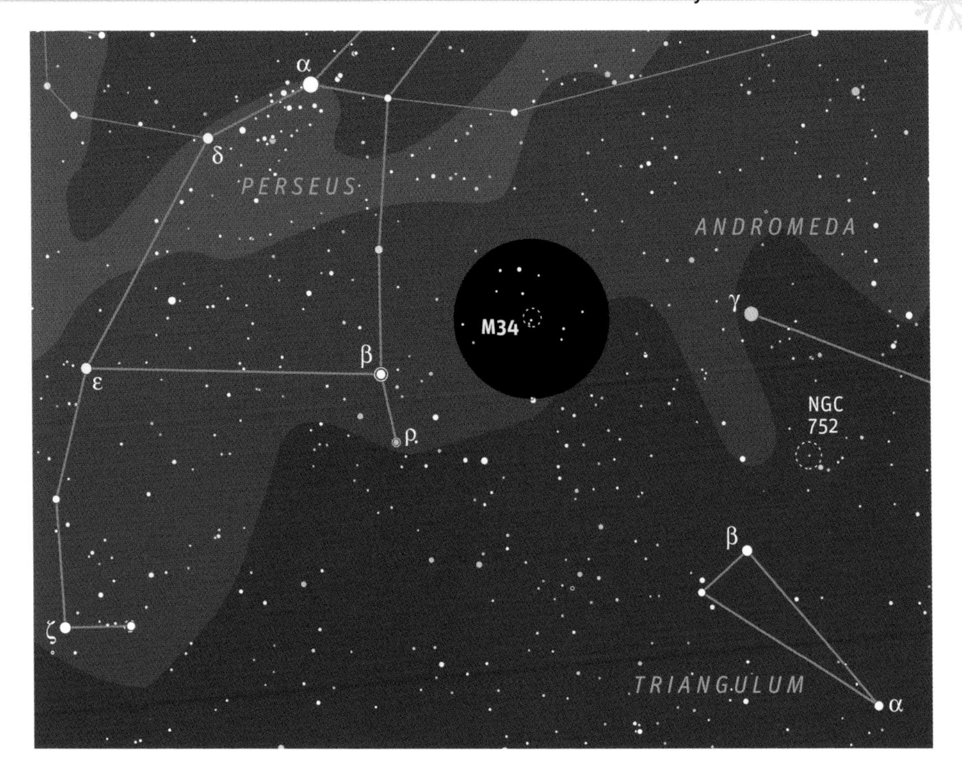

M34: A True Binocular Cluster

For binocular observers under the glow of light pollution, the list of deep-sky objects that display significant detail can seem discouragingly short. All but a handful of galaxies and nebulae require fairly dark skies to be seen at all, and even relatively high-surface-brightness targets like globular clusters lose much of their appeal under bright skies. However, one class of object does stand up reasonably well under adverse sky conditions — open clusters. And one that fares better than most is M34 in Perseus.

In June 2000, I came to appreciate how resilient this particular cluster is. In the early hours of a sweltering summer morning, I got up to look at Comet LINEAR as it slipped by M34. In spite of the cluster's low altitude and the brightness of my suburban backyard, M34 looked great. The comet didn't fare so well, but the cluster's brightest dozen stars were easy pickings in my 10 × 50s. Under better conditions, these stars are joined by a scattering of fainter cluster members.

To find M34 simply sweep slightly north of a line joining Algol and the pretty golden yellow star Gamma (γ) Andromedae.

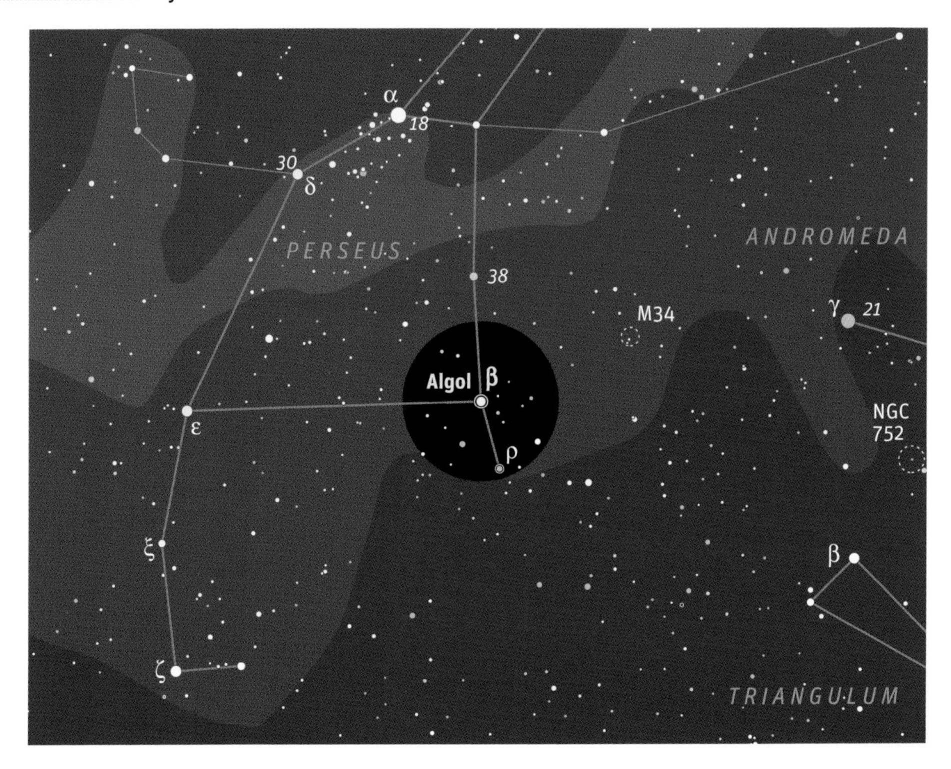

Watching the Demon Star

Among the most exciting astronomical events are eclipses of the Sun. Anyone who witnesses the Moon's disk silently extinguishing the Sun's light can never forget it. Unfortunately, to see one you usually have to travel a great distance. But if you turn attention to the night sky, chances are you can witness a stellar eclipse this week from your own backyard.

Algol (β Persei), sometimes called the Demon Star, is the archetypal eclipsing binary. Every 2 days 20 hours 49 minutes, the primary star is eclipsed by its dimmer companion, and the system fades from magnitude 2.1 down to 3.4. Not only are the eclipses frequent and widely visible; unlike solar eclipses, which speed by in a matter of a few short minutes, these last 10 hours! Granted, Algol is not as spectacular as a total eclipse of the Sun, but watching eclipsing binaries wink is a fascinating activity.

Algol and some convenient comparison stars are of naked-eye brightness and easy to find. (In the chart above, star magnitudes are given with their decimal points omitted, so the star labeled 38, for example is magnitude 3.8) The star itself lies in an attractive binocular field and forms the northeastern corner of a 2°-wide asterism that resembles a fainter, distorted version of the Bowl of the Big Dipper. Its southeastern star is reddish. At its brightest Algol is a good match for Gamma (γ) Andromedae.

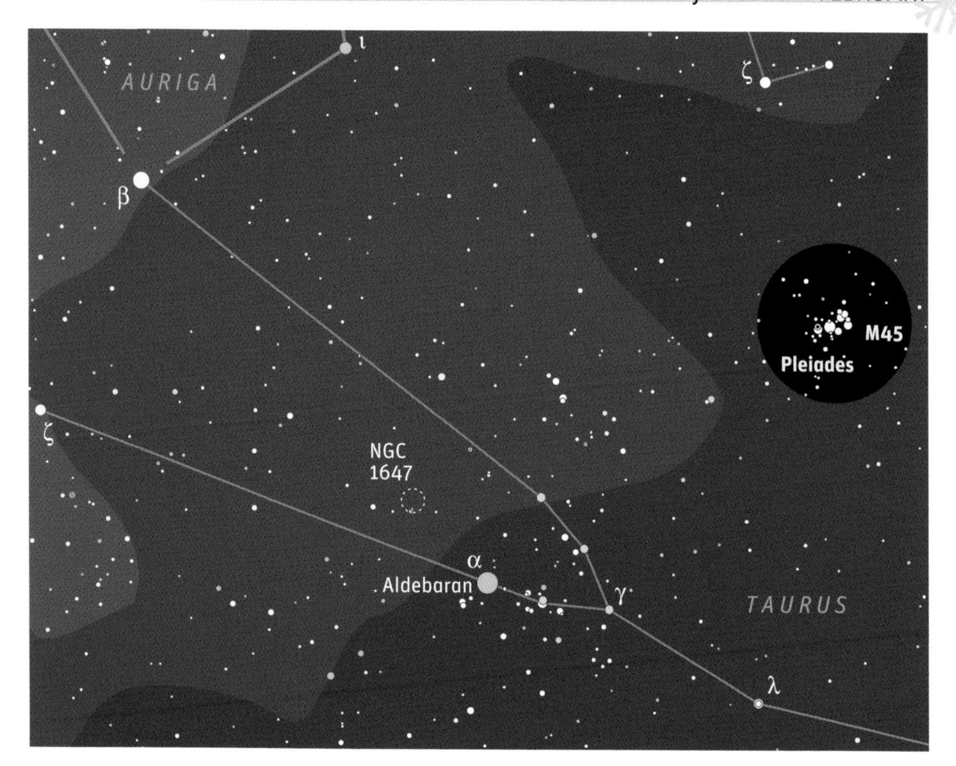

The Pleiades

Any list of the most spectacular binocular fields in the entire sky will surely include the Pleiades near the top. All the descriptive superlatives were used up long ago. "Spectacular," "breathtaking," "stunning" — they all fit, but they sound like clichés. What these well-worn phrases can never truly convey is the actual beauty of the cluster as seen through everyday binoculars. It is a view that one must experience firsthand to fully appreciate.

The cluster's five brightest stars are arranged in a "little dipper" configuration. But what makes the view extra special to me is all the fainter stars sprinkled about the cluster's leading lights. Particularly delightful is the curving row of five 7th-magnitude stars beneath the dipper's handle and the tight, very difficult 8th-magnitude double star, South 437, in the middle of the bowl.

The Pleiades rank with Saturn's rings and the craters of the Moon as sky objects that can spark a lifelong interest in astronomy. As Leslie Peltier recounted in his classic autobiography, *Starlight Nights,* it was a childhood gaze at this cluster out his kitchen window in 1905 that started him on the starlit path he followed the rest of his life. Perhaps a look at the Pleiades will inspire you as well.

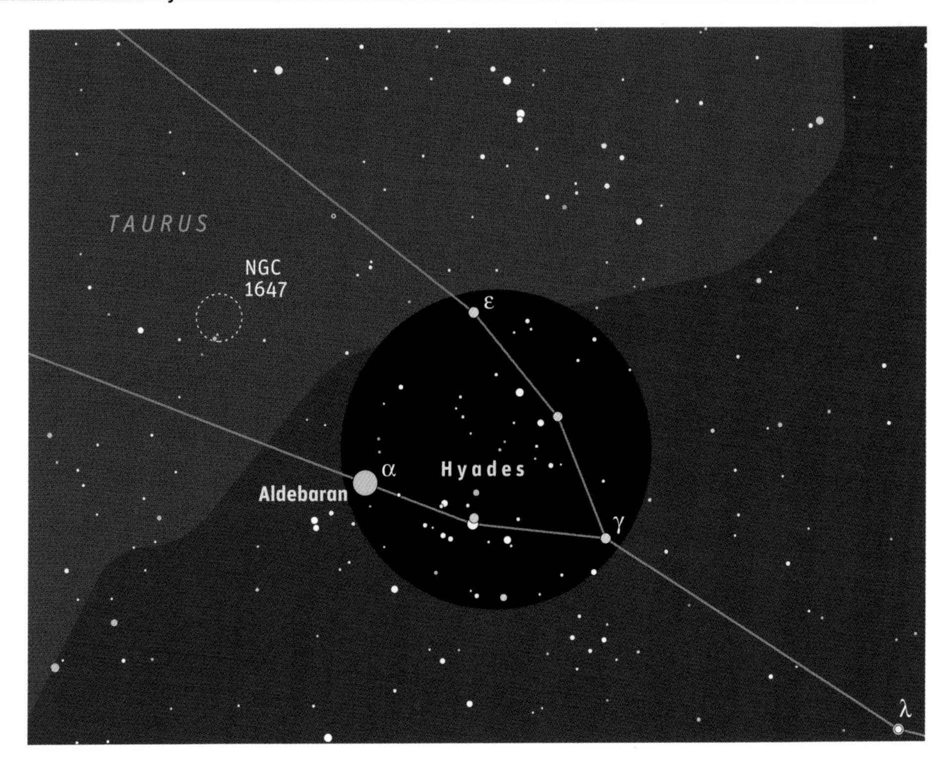

Corralling the Hyades

Although beginning observers often assume that telescopes offer the best views of the night sky, a number of objects actually look much better in ordinary binoculars. No doubt about it, when it comes to light gathering and resolution, any decent-size telescope is going to produce brighter and more detailed views of small objects than common 10 × 50 or 7 × 50 binoculars. But when it comes to showing the most sky in a single view, binoculars come out on top. This is never more true than for corralling a really big star cluster, such as the Hyades in Taurus.

The Hyades span about 6° of sky because they constitute one of the nearest open clusters, being only 150 light-years away. Most telescopes will show a very sparse, uninspiring sprinkle of stars spilling out of the view. But binoculars reveal the whole thing: a big field full of suns arranged in interesting miniconstellations and geometric shapes.

Brightest is orange-hued Aldebaran, but its distance (65 light-years) is one giveaway that it is not actually a member of the Hyades. However, its presence adds so much to the appeal of the field of view that observers can justifiably consider it an honorary cluster member.

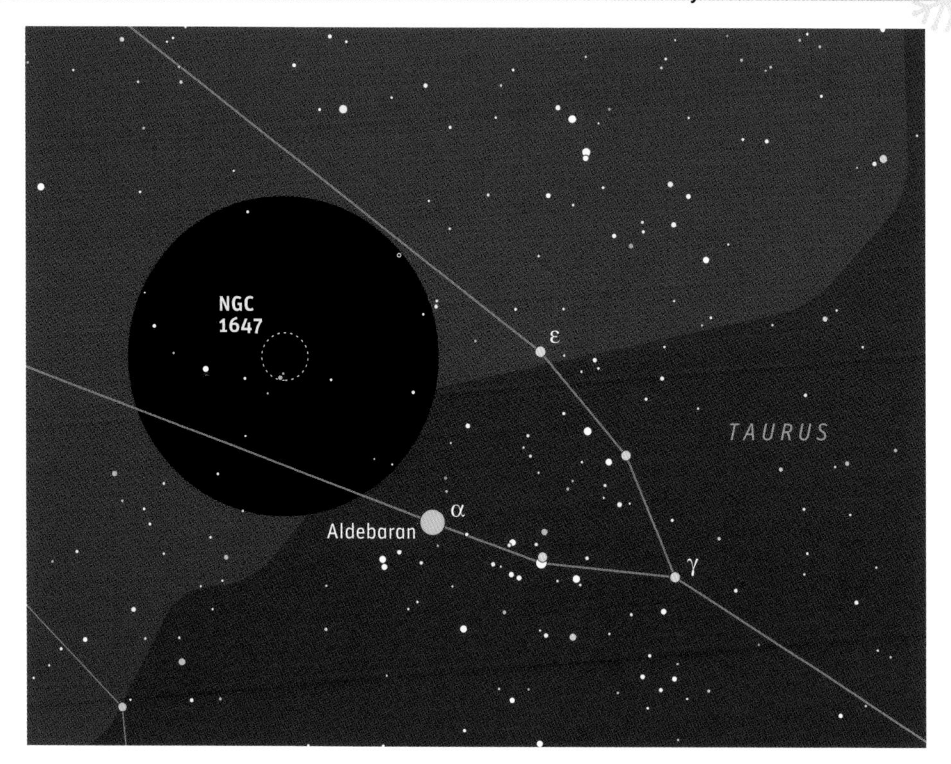

NGC 1647: The Crab Cluster

With spectacular binocular highlights like the Pleiades and the Hyades so close by, it's little wonder that Taurus's lesser clusters go unnoticed. In the case of NGC 1647, that's too bad. Were it almost anywhere else in the sky, it doubtless would receive more attention.

This sprawling cluster is found 3½° (about a half binocular field) northeast of bright Aldebaran. You'll need a moonless night and transparent, dark skies to see it at its best, since most of NGC 1647's stars are fainter than 9th magnitude.

In my image-stabilized 15 × 45 binoculars, the cluster looks like a crab — two curving rows of stars toward the northwest and west suggest the pincers, four brighter outliers mark the tips of legs, and the clump of stars in the middle form the crab's body. Perhaps it was because I observed NGC 1647 from the seashore that the image of a celestial crustacean came so readily to mind. Have a look and see what you think. If at first the crab shape doesn't present itself, try observing the cluster again from a beach and see if that helps fire your imagination!

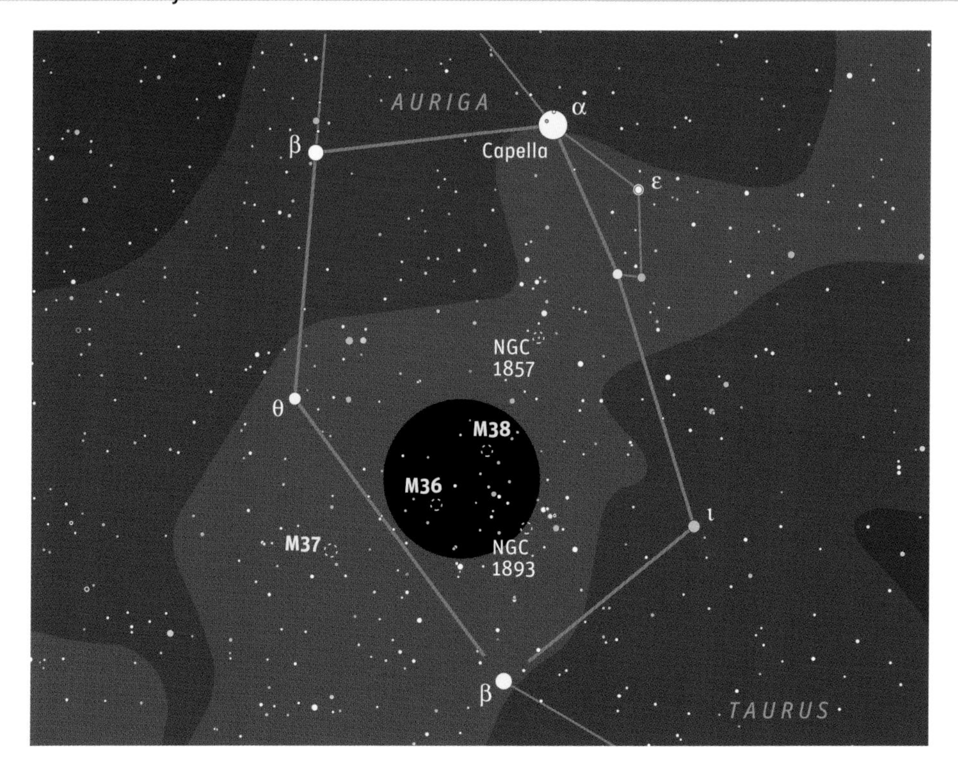

Auriga's Messier Clusters

Some binocular objects are impressive on their own, while others are memorable because they are situated in a remarkable region of sky. Then there's the third category — impressive objects lying in remarkable fields. That's the case for the trio of Messier open clusters in Auriga: M36, M37, and M38. Each could be the subject of its own binocular highlight, but here the whole really is greater than the sum of its parts.

One thing that makes the three so enjoyable is that their proximity invites comparisons. Indeed, depending on your binoculars, you might just be able to fit all of them into the same field of view.

Observing with 10 × 30 image-stabilized binoculars from my light-polluted backyard, I find M36 the easiest of the three to see. This is partly because it has the smallest apparent size and a high surface brightness. It has a distinctive spiderlike appearance, with rows of barely perceived individual stars radiating from the cluster's center.

M38, on the other hand, shows no distinct character — it is simply a large, diffuse glow with a few faint stars occasionally adding a random sparkle. Because of this it suffers the most from the adverse effects of a bright sky, and under unfavorable conditions it can be difficult to see at all.

As for M37, its appearance lies somewhere between those of its neighbors — not as diffuse as M38, and not as conspicuous as M36.

Give these clusters a try and see if your impressions match mine. But don't forget also to appreciate them as a group. Lying on a rich Milky Way background, the Auriga trio make up one of the most rewarding binocular regions in the entire Northern Hemisphere sky.

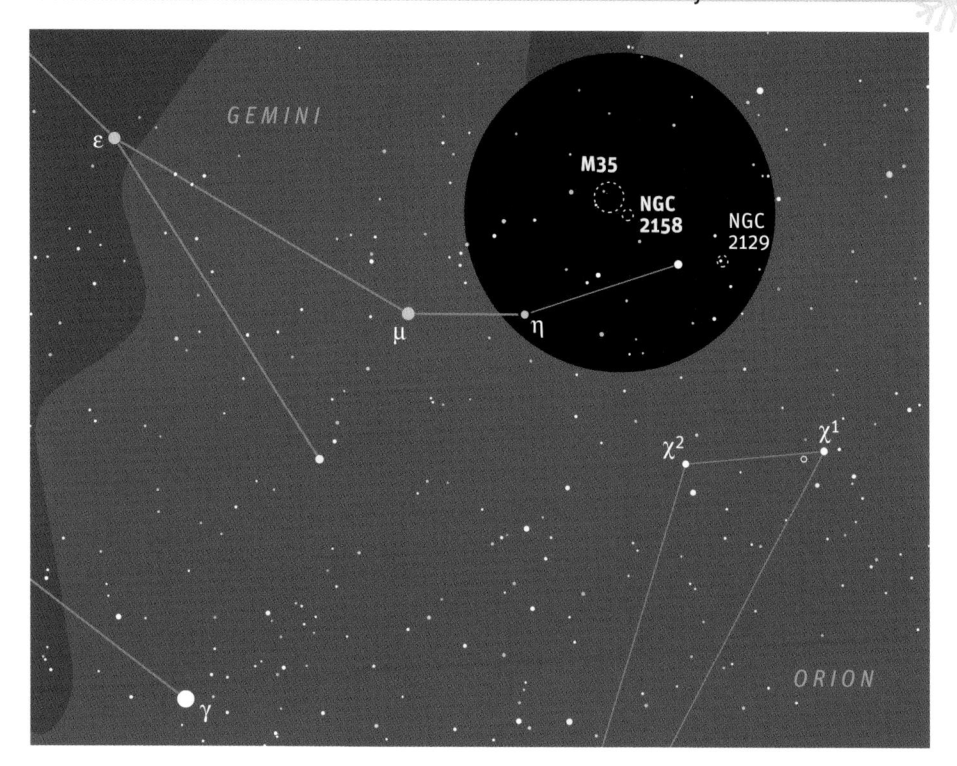

Another Double Cluster

One of the finest binocular objects is the famous Double Cluster in Perseus. But a lesser-known double cluster in Gemini also lies high on winter evenings. Compared with the Perseus pairing of twin, rich clusters, the Gemini clusters M35 and NGC 2158 are a celestial odd couple. M35 is a bright and easy binocular target, while NGC 2158 is faint and challenging. The two are easy to locate near Gemini's westernmost foot.

M35 is Gemini's most rewarding binocular sight. I can see a half dozen individual stars sparkling through the sodium-vapor pall of my suburban night sky. These bright stars form an east-west band set against the larger, round glow produced by dozens of fainter stars at the verge of resolution. NGC 2158 is a different kettle of fish. Here the challenge is to see the cluster at all! To succeed you will need a dark, moonless sky, a steady hand (or better yet, some kind of mount or image-stabilized binos), and a magnification of at least 10×. Even then, NGC 2158 will appear as little more than a very dim, small, hazy glow situated on the southwestern edge of M35. Like most binocular sights, more magnification produces better views — particularly through light-polluted skies.

To fully appreciate the view, keep in mind that you are seeing the effect of distance. M35 is about 3,000 light-years away, while NGC 2158 lurks in the background, five or six times farther away. Little wonder that NGC 2158 is a mere ghost of M35.

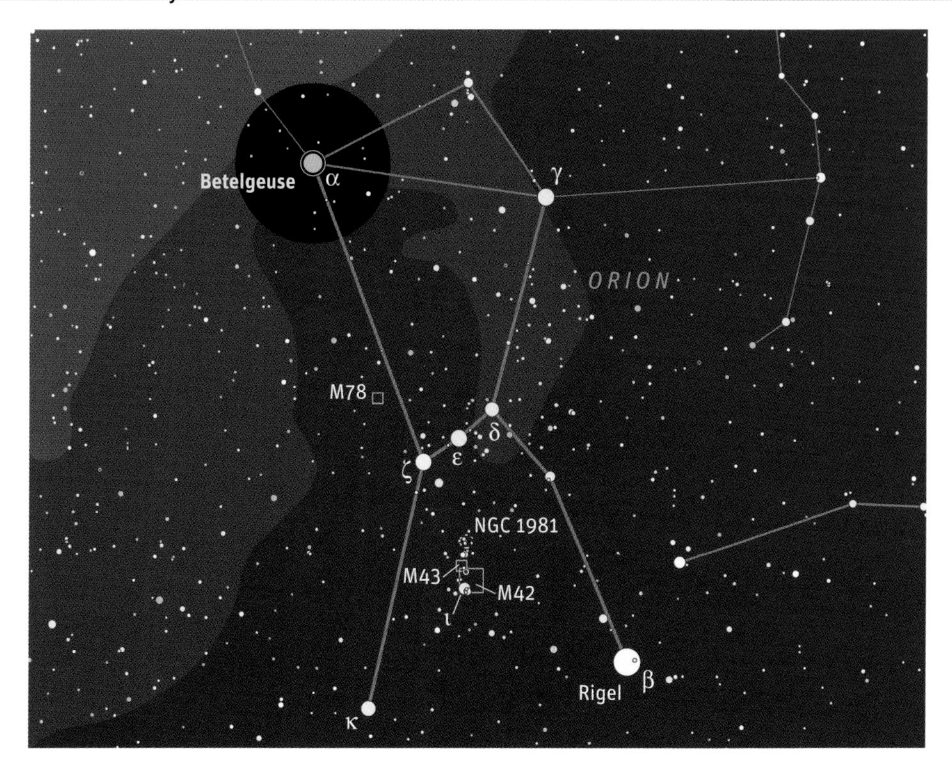

Golden Betelgeuse

We have all read descriptions of stars appearing "blood red" or "vivid blue," but in reality most star colors are subtle. The well-known "red supergiant" Betelgeuse (α Orionis) is typical. Betelgeuse is a type-*M* star, which means that most of its visible light is in the yellow, orange, and red parts of the spectrum. But it also pumps out enough light at shorter wavelengths that its dominant color is diluted. The result is a star that is not so much strongly colored as lightly tinted.

One factor that influences how well star color shows up is the amount of light that reaches our eyes — and here's where binoculars help. To the naked eye, Betelgeuse has a pale tint that is easiest to discern if you look back and forth between it and its bright neighbor, icy white Rigel. However, in binoculars Betelgeuse has a strikingly beautiful golden orange hue. Why is that?

Your eyes have two kinds of detectors: *rods*, which are sensitive to faint light, and *cones*, which allow us to see color but are relatively insensitive. In dim conditions we mostly use our rods, so most faint stars appear colorless. The light-gathering power of binoculars gives the cones enough light to perceive stellar hues better.

But colors are also washed out if you see *too much* light. Try this experiment with Betelgeuse: defocus your binoculars slightly so Betelgeuse expands from a sharp point of light into a somewhat fainter disk. Doing this reduces any overexposure of Betelgeuse's tiny point image on your retina. Overexposure makes any color look paler (more white) than it really is.

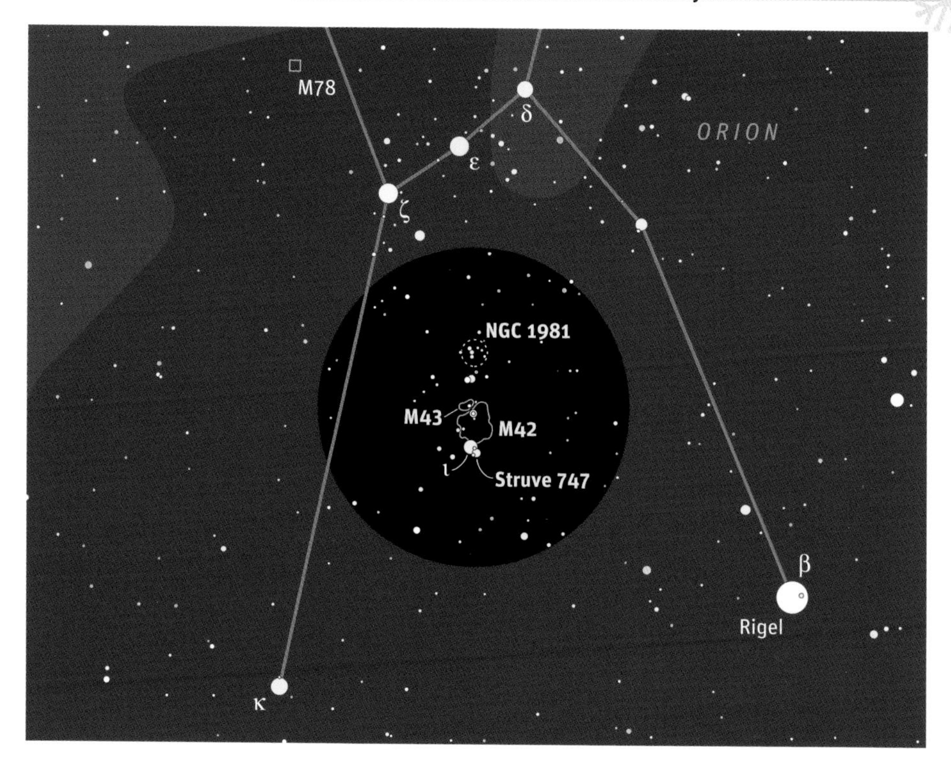

Orion's Sword

Few binocular sights pack as great a visual impact as Orion's Sword. Who could look at this collection of glittering stars and glowing nebulosity and fail to be impressed? Granted, nothing beats the sight of the Orion Nebula, M42, in a good telescope, but the binocular view provides a context — showing not only the nebula itself, but the entire neighborhood it calls home. And an attractive home it is indeed!

Orion's Sword is really three highlights in one. Of course the center of attention is M42 itself. Even under poor conditions, this nebular mist can be seen enshrouding three points of light. These are the Trapezium (its four main stars reduced to a single 5th-magnitude blip by the low magnification of binoculars), 5th-magnitude θ^2 (Theta^2) Orionis, and its 6th-magnitude neighbor to the east. Together they are an arresting sight worthy of the accolades heaped upon them.

Due south of M42 lies Iota (ι) Orionis. At magnitude 2.8 it is the brightest star in the field of view, but look carefully at the star lying 8′ southwest. Notice anything? This is the double star Struve 747. In my 10 × 50 binoculars I can just split this pair of 4.8- and 5.7-magnitude suns — but only when I use a tripod to steady the view. My 15 × 45 image-stabilized binos have an easy time resolving the double.

The northernmost attraction in Orion's Sword is the loose open cluster NGC 1981. Even under bright suburban skies, steadily held 10× binoculars show a handful of cluster stars. Although this grouping is often overlooked because of its showier neighbor, it's an attractive cluster worth a long, careful look.

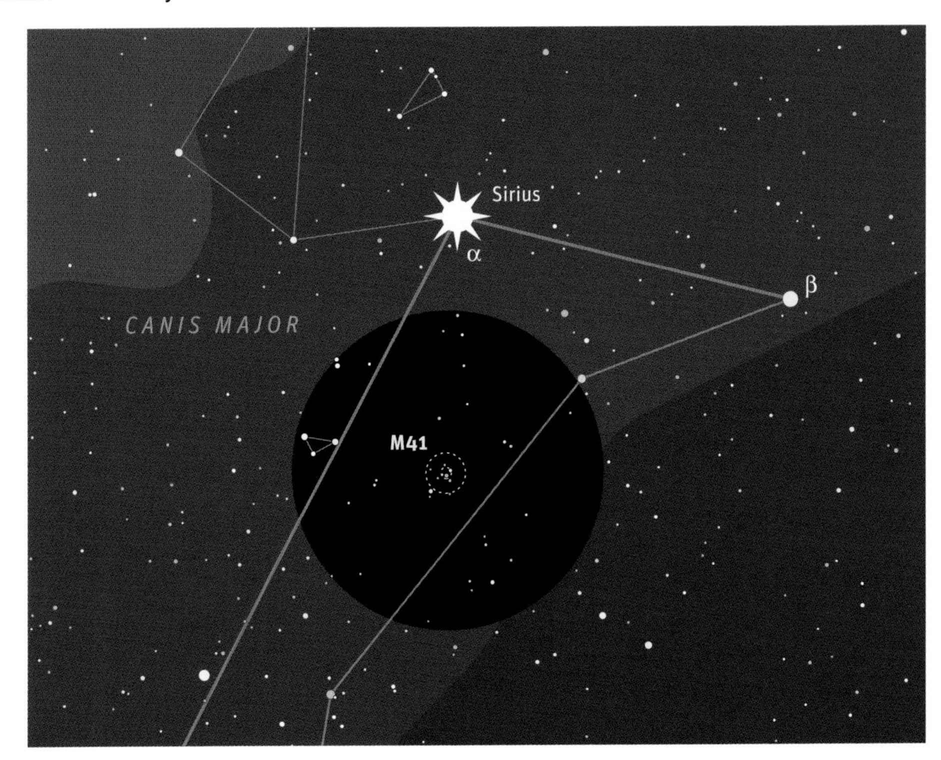

The Overlooked Open Cluster: M41

Few stars are as alluring as Canis Major's leading light, Sirius. When you're out for an after-dinner walk on a crisp winter's evening, dazzling Sirius draws the eye skyward. For those who cannot resist the temptation to take a look with binoculars, a surprise awaits near the bottom of the field of view: the open cluster M41. I wonder how many times this delightful clump of stars has been "discovered" in this fashion by novice stargazers. Certainly, having a brilliant marker like Sirius makes it easy to locate. In his book *The Messier Objects,* Stephen James O'Meara captures the scene beautifully: "As Canis Major, the Great Dog of Orion, rises above cool winter landscapes, open cluster M41 hangs below its collar like an ice-covered tag reflecting moonlight."

M41 is thought to lie some 2,300 light-years away. Even with strong moonlight (or city light pollution) the half dozen brightest stars of this "ice-covered tag" still shine through. Note the general richness of the field — a result of Canis Major's position in the winter Milky Way. Particularly eye-catching is a nearby pair of triangles (indicated), one a half field north of Sirius and the other a field south-southeast. Although lacking the density of riches found in summer, the winter Milky Way is home to many strikingly beautiful star fields, as you can discover for yourself with a little binocular exploration on your next winter-evening stroll.

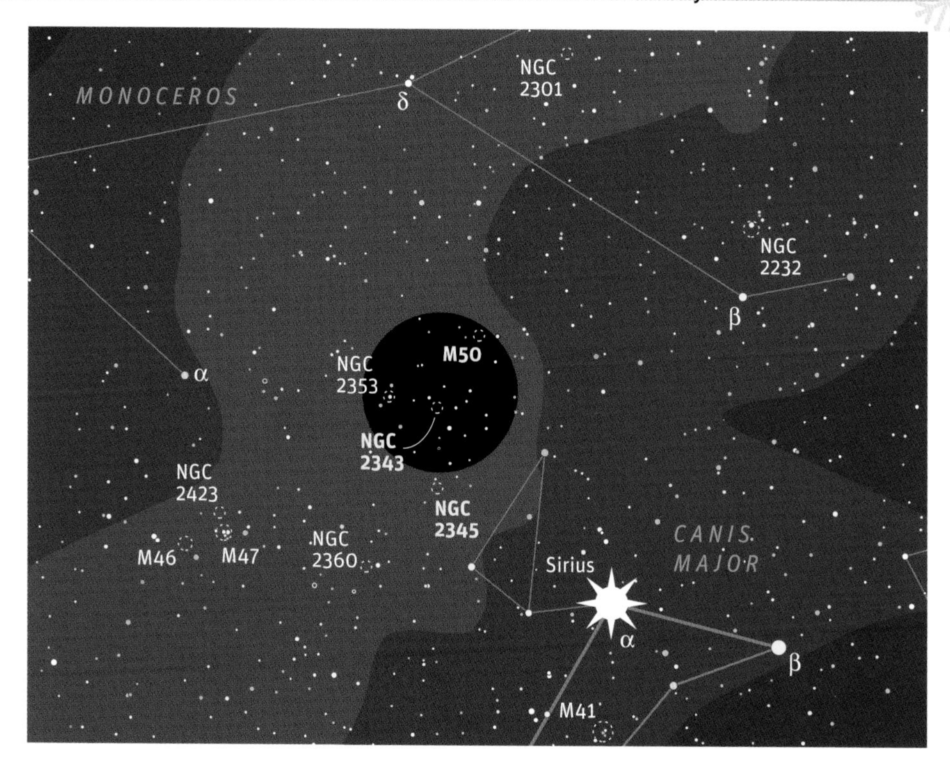

M50 in a Stream of Starlight

The broad and bright summer Milky Way has often been likened to a river of light. By comparison, the winter version is a shallow stream, but like its warm-weather counterpart, it too is worthy of leisurely fishing expeditions to catch some of its lesser-known wonders. Particularly rich and rewarding is the stretch of starlight shown in the map above, extending from the star clusters Messier 46 and 47 in Puppis northwestward into Monoceros.

M50 is the best known of the open clusters lying north-northeast of brilliant Sirius. My 10 × 30 image-stabilized binoculars show it as a distinct little elongated glow in an attractive field peppered with 7th-magnitude stars. The view from a sky darker than my light-polluted suburban backyard is even more rewarding. Still, one can win back some of what the bright sky takes away by using higher-magnification binoculars. And sure enough, my 15 × 45s do a much better job of showing the cluster, even revealing a smattering of individual stars. But of course, for binoculars of this magnification to show more detail they need to be stabilized, either electronically (as mine are) or by some kind of support.

The benefits of extra magnification are even more pronounced with M50's neighbors. Although NGC 2343 is an easy find in the 15 × 45s, it can be only glimpsed with the 10 × 30s. NGC 2345 is a difficult object in the bigger binoculars and, unsurprisingly, doesn't show at all in the small ones. That said, it wouldn't surprise me to find that both objects are relatively easy under dark skies. And if you grow weary of the challenge of fishing out faint clusters, you can always slip a ways downstream (and over to the next page) and enjoy the splendid and bright M47.

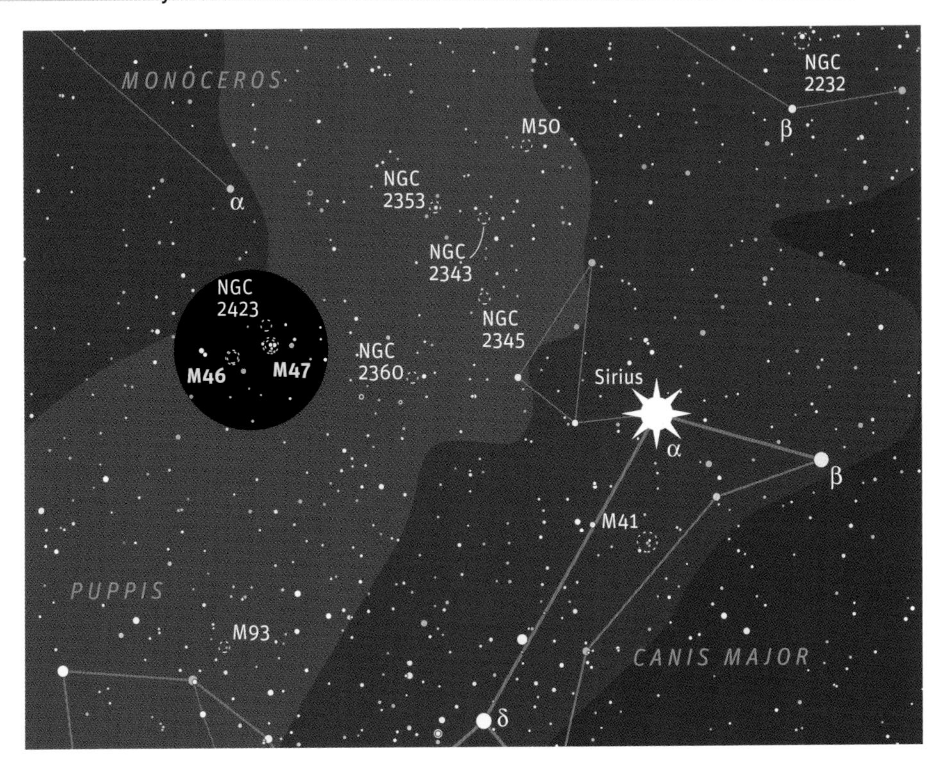

M46 and M47: A Celestial Odd Couple

Most everyone is familiar with the Double Cluster in Perseus. As noted on page 17, this pairing is one of the sky's best binocular sights. However, this season also offers a pair of Messier clusters: M46 and M47 in Puppis.

Although similar in size and overall brightness and separated by only 1½°, the clusters in this pair look very different from each other. In his book *The Messier Objects,* Stephen James O'Meara quips that their dissimilarity is "like trying to compare a flower with a rock."

Of the two, the westernmost, M47, is the more readily seen. At its core is a grouping of a half dozen bright stars arranged in a pattern that suggests the constellation Sagitta, the Arrow. These bright stars make M47 easy to find even in moonlit or light-polluted skies.

By contrast, M46 is all but impossible to see in unfavorable conditions. From my suburban yard I could glimpse only the barest hint of it even in 15 × 45 image-stabilized binoculars. This is not surprising because, though very rich with stars, M46 contains none brighter than magnitude 9. Under dark country skies you will see, as O'Meara did, "a round, uniform 6th-magnitude glow."

Two Southern Delights

Binoculars offer the ultimate in grab-and-go viewing. Once in a while it is refreshing to leave the telescope behind and head outdoors for a quick look at something of special interest, or just to casually sweep the sky. During one such random walk in the winter Milky Way, I chanced upon a lovely pair of open clusters scraping the southern horizon in the sprawling constellation of Puppis.

Situated in the same field of view as 2nd-magnitude Zeta (ζ) Puppis is NGC 2477. Under dark skies ordinary binoculars show the cluster as a dim, round haze. This is an unusually rich galactic cluster packed with faint stars — which is why it doesn't look like a typical open cluster.

For something that better matches expectations, look 1½° northwest for neighboring NGC 2451. You should be able to make out a dozen or so stars huddled around a lovely orange 3.6-magnitude jewel.

To see NGC 2477 and 2451 you'll have to find an unobstructed southern horizon and choose a time when they are at their highest due south.

SPRING

Akira Fujii

2

MARCH • APRIL • MAY

PLANETARY NEBULA
GLOBULAR CLUSTER
DIFFUSE NEBULA
OPEN CLUSTER
VARIABLE STAR
GALAXY

ABOUT THE CHARTS:

Each of the star maps in this chapter has been rendered at one of three different scales: the wide-field charts to magnitude 7.5, the medium-scale charts to magnitude 8.0, and the close-up charts to magnitude 8.5. Regardless, the darkened circular area always represents the field of view for typical 10 x 50 binoculars.

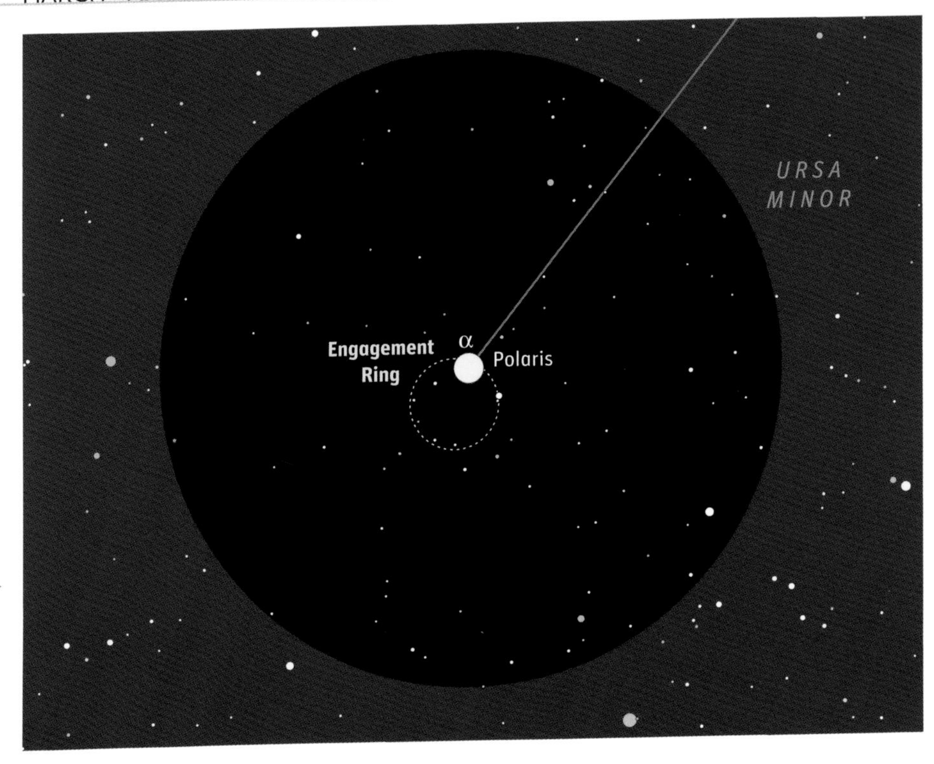

The Engagement Ring

The sky is full of interesting little asterisms — especially if we go hunting for them with binoculars. Most everyone has heard of the Coathanger in Vulpecula (page 64), but there are lots more, and no doubt plenty await discovery. Most of these stellar groupings are not gravitationally connected, but the ability of the eye and brain to construct patterns out of randomly distributed points can't be denied. After all, that's exactly what constellations are.

One of the more engaging (if you'll forgive the phrase) asterisms is the Engagement Ring in the Little Dipper. This stellar ring is complete with a sparkling 2nd-magnitude diamond, Polaris. The ring itself is a 35-arcminute-diameter ragged circlet of mostly 8th- and 9th-magnitude stars on the side of Polaris opposite the Little Dipper's bowl. Depending on your sky conditions, just about any size binoculars should show this grouping.

Because of its proximity to the north celestial pole, the Engagement Ring is visible all night long, every night of the year, for Northern Hemisphere observers. Have a look for yourself.

The Power of Power

Riding high in the north are two of the Messier catalog's most conspicuous galaxies: M81 and M82. This pair offer a fine demonstration of the power of power. Under suburban skies, I viewed this galaxy duo with three pairs of binoculars, each offering a different magnification. (As explained on page 6, the magnifying power is the first number in a binocular's specification. For example, 10 × 50 binoculars have a power of 10×.) With the 7 × 50 pair, only M81, the brighter of the two galaxies, could be seen. The extra magnification of the 10 × 50 binoculars showed M81 easily while M82 could be glimpsed only part of the time. By far the best view was with the 15 × 45 binoculars. Not only were both galaxies easily seen, but I could also tell which was which based upon size and orientation — in spite of the fact that with 45-millimeter objectives, these binoculars gathered a little less light than the other two! Higher magnification results in a larger image scale, improving the visibility of low-contrast objects. This makes hunting faint targets a little easier — regardless of sky conditions.

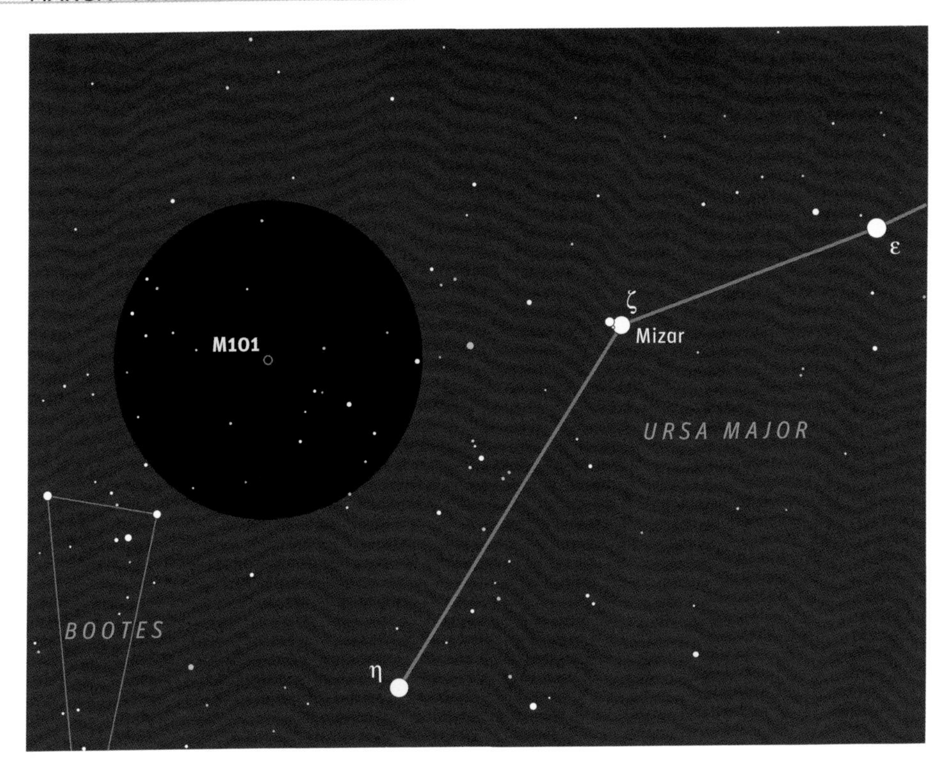

M101 in the Mind's Eye

A great deal of the enjoyment we derive from viewing the night sky comes from within. The face-on spiral galaxy M101 in Ursa Major is a case in point. The galaxy's location is easy to find by following the ragged trail of 5th-magnitude stars that runs northeast from the famed double star Mizar, the middle star in the Big Dipper's Handle. Under a dark sky, M101 appears as a small, round, faint fuzz of light in my 10 × 30 image-stabilized binoculars. Under less favorable conditions, such as my light-polluted suburban backyard, it can be difficult to see at all even with larger binoculars. But to simply note its appearance rather misses the point — and this is true for countless other objects, whether viewed in binoculars or in telescopes.

Pure visual spectacle is only one reason to hunt down deep-sky targets; the other is the intellectual enjoyment of contemplating what you are viewing. So consider this the next time you visit M101 with your binoculars: that faint little smudge represents the combined light of several hundred billion suns whose luminance is reduced to a feeble glow by a distance of approximately 22 million light-years. Or turn the situation on its head and consider that from a planet orbiting a star in M101, our own, smaller Milky Way Galaxy would present an even less impressive sight. Our Sun shines so brightly in the daylight sky that it's difficult to make the mental leap required to grasp how any distance could be great enough to diminish its brilliance (and that of billions of its Milky Way cousins) so fantastically. But this gives us some small insight into just how far 22 million light-years really is. It's all a matter of perspective.

A Star-Hop to M51

For binocular observers, galaxies are among the most challenging deep-sky objects — the vast majority are small and faint. But a few are small and reasonably bright. The trick to locating these galaxies is careful star-hopping — working step by step from a known starting point to the precise location of your quarry, in a careful and deliberate fashion.

Located below the handle of the Big Dipper, M51 is a good galaxy on which to try out your star-hopping skills. Begin by planning your route on a detailed star map, like the one above. Start with an easy-to-find star, such as Eta (η) Ursae Majoris, at the end of the Dipper's handle. Next, make patterns (triangles, lines, etc.) out of the field stars to guide you the rest of the way. For example, I always see Eta as the first star in an arc of three that spans about one binocular field. Then I look for a small triangle of stars below the middle star of the arc. Once I locate that triangle, all I need remember is M51's position relative to it. Try this route yourself. If you star-hop carefully, the galaxy should be visible as a little smudge of light — your reward for a well-executed star-hop.

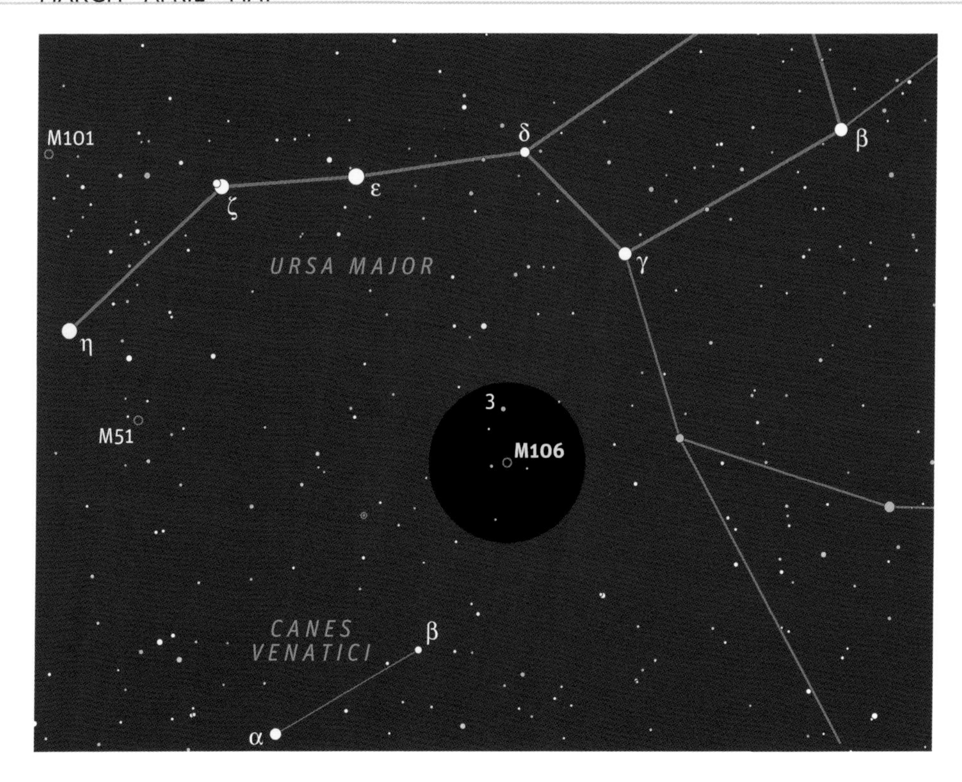

Galaxy Season

In the Northern Hemisphere, spring is galaxy season. With the Milky Way clinging to the horizon, we enjoy an unobstructed view of the universe beyond our own celestial backyard. The swath of sky running from below bright Spica in the south all the way north to Draco is brimming with distant island universes. For viewers at midnorthern latitudes, one of the finest examples is nearly directly overhead in the evening: M106. This spiral galaxy is situated below the handle of the Big Dipper, in Canes Venatici — 22 million light-years beyond the stars that make up this small constellation.

To find M106, aim your binoculars midway between Beta (β) Canum Venaticorum and Gamma (γ) Ursae Majoris. The galaxy lies 1.7° due south of the 5th-magnitude orange star 3 Canum Venaticorum, in the same field of view. A 6th-magnitude star is just ½° to the galaxy's east. Under a dark sky almost any binoculars should show M106 as a small smudge. Observers with less optimal conditions may need at least 50-mm binoculars that magnify 10× or more to glimpse one of the galaxy season's finest catches.

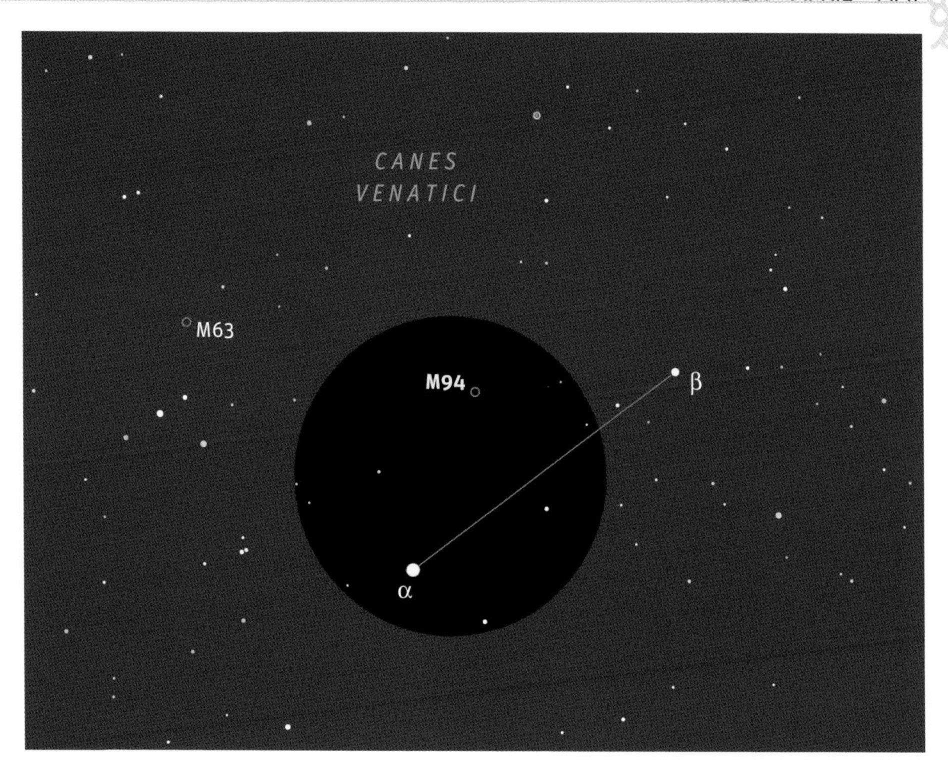

Steady on Galaxy M94

Recently I enjoyed a little galaxy observing with my telescope, which is equipped with a standard 7 × 50 finderscope. I swung the scope over to Alpha (α) Canum Venaticorum and looked in the finder. My target, the galaxy M94 (magnitude 8.2), was plainly visible 3° northwest of the 3rd-magnitude star, in spite of my light-polluted sky. I was surprised at how easy the galaxy was to locate, because a few nights before I had hunted for it with my 10 × 50 binoculars and had found it considerably more difficult. Could sky conditions be *that* much better this night? Curious, I dug out the binoculars for another look and again was faced with a challenging find. What was going on?

Because the binoculars yield higher magnification and allow the use of two eyes instead of one, the view should be better than in the finderscope. But the finderscope has one big advantage: it is firmly attached to a well-mounted telescope, while the binoculars are hand-held. Yes, a steady view is *that* important.

There are a number of ways to give your binoculars shake-free support. Camera tripods and dedicated binocular mounts are often used with great success. Leaning the binoculars against a fence post or some other structure also helps. By now you will have noticed that I do much of my observing with image-stabilized binoculars. To me, these represent the best of all possible worlds — steady views without the encumbrance of extra equipment.

The next time it's clear, set your sights on M94 and on 8.6-magnitude M63 nearby — and see for yourself how important a steady view is. Try viewing the galaxies with and without a support. I think you'll notice a big difference.

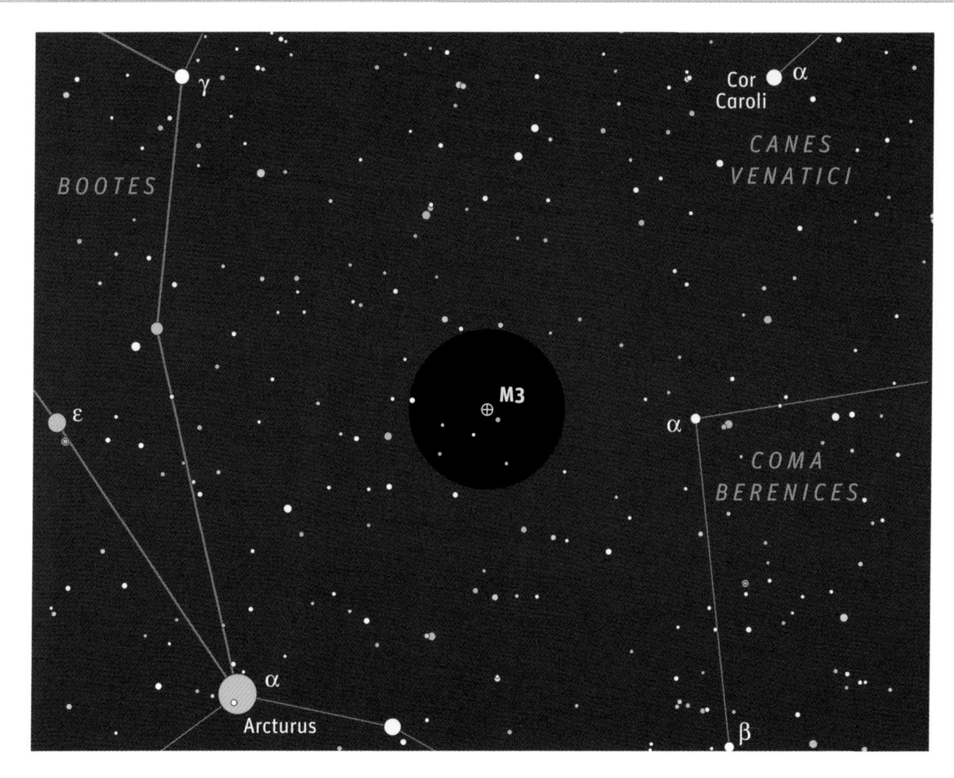

Globular Cluster M3

Nature lovers note the change of seasons not just with calendars but by observing a variety of repeating cycles. For me, the end of spring in the Northern Hemisphere seems at hand when the galaxy fields of Leo, Virgo, and Coma Berenices begin to relinquish the sky overhead to the bright globular clusters I associate with summer. An early forerunner of this flock is M3, which is already very high in the east on April and May evenings.

M3 is most easily found with binoculars by scanning a little less than halfway from Arcturus to Cor Caroli, or Alpha (α) Canum Venaticorum, which lies beneath the Big Dipper's handle. Glowing dimly at 6th magnitude, M3 lies less than ½° northeast of a star of similar brightness.

If your binoculars are properly focused, M3 will appear noticeably fuzzy, in sharp contrast to its neighbor. The higher your magnification, the more obvious the difference between the two becomes. I usually start my sweep for M3 from brilliant Arcturus, which is probably why I always think of this globular as belonging to the constellation Boötes instead of where it actually resides, over the border in Canes Venatici.

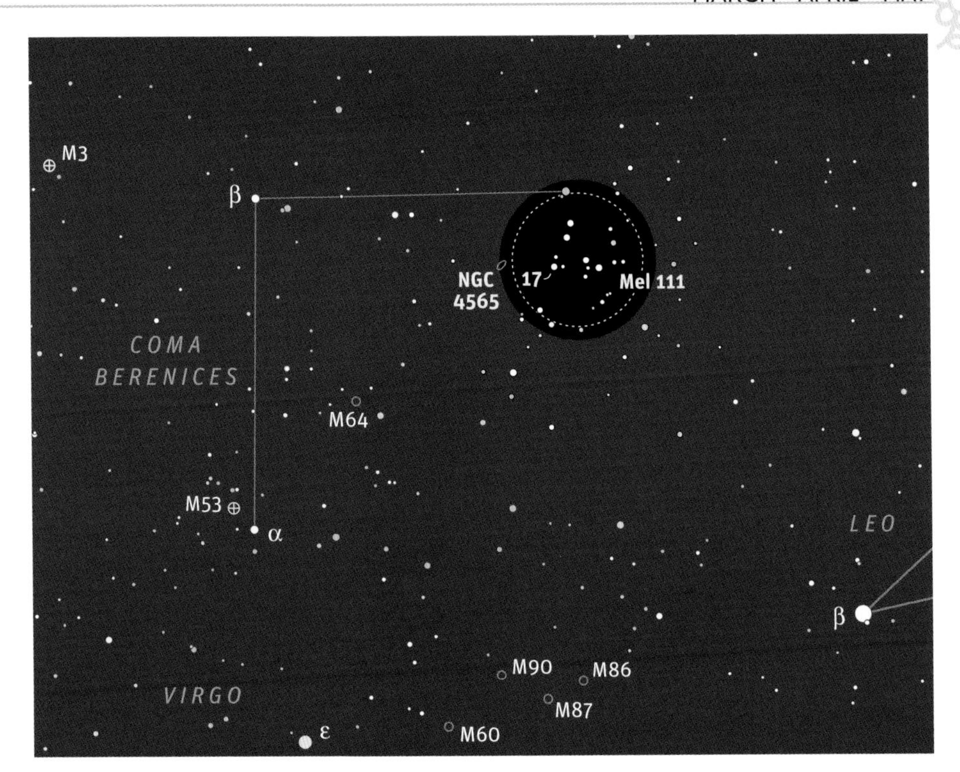

In and Around Melotte 111

Open clusters range from tiny smudges to bold splashes of starlight spanning several degrees of sky. One of the latter is Melotte 111, which constitutes much of the constellation Coma Berenices. The cluster's apparent size is the result of its proximity — Mel 111 is one of the nearest open clusters, residing only 300 light-years away. To the naked eye on a dark night, it looks the way typical clusters look in a telescope. In binoculars it fills the field of view with two dozen stars brighter than 8th magnitude.

Considering the rather sparse nature of the spring stars, a binocular field this rich is something to linger over and enjoy. Notice the pretty double star 17 Comae Berenices. Its 5.3- and 6.6-magnitude components are separated by 145″ — an easy split for any binocular.

Mel 111 lies right in the heart of spring galaxy country and if your skies are dark, you might just glimpse one or two lurking in the background. Try your luck with NGC 4565, which lies 1½° east of 17 Comae Berenices. At 10th magnitude it is the brightest galaxy near the cluster.

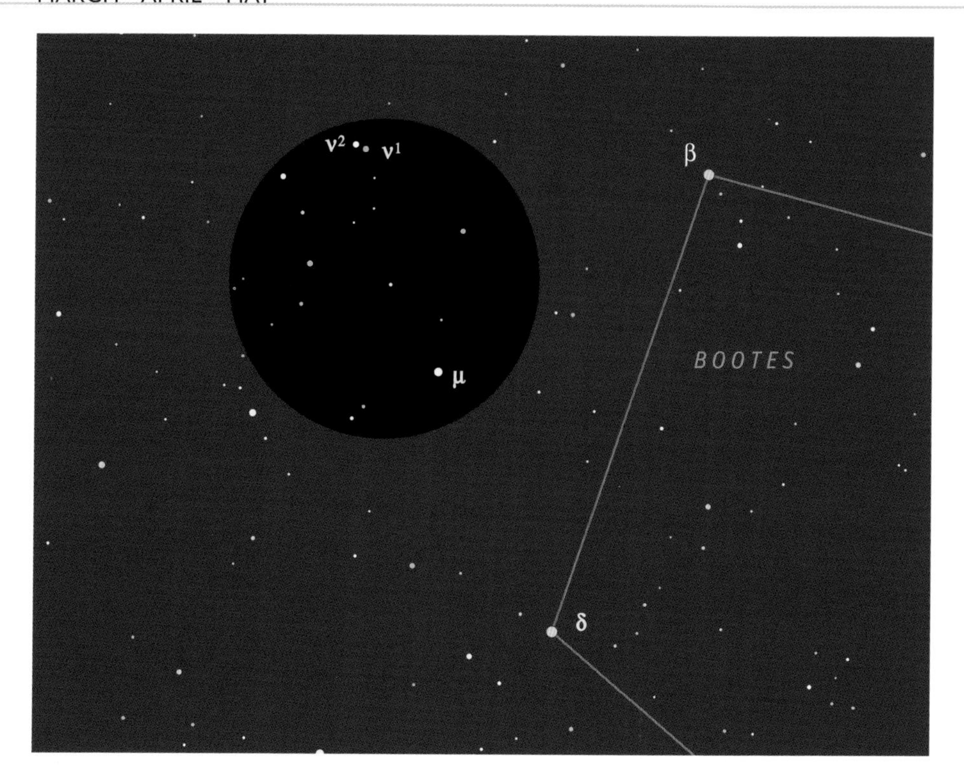

A Trio of Doubles in Boötes

Boötes isn't a constellation that binocular observers spend a lot of time exploring. That's not entirely surprising when you consider that out of the 109 Messier objects, exactly zero call this constellation home. So, if Boötes holds no bright clusters, galaxies, or nebulae, what is there to see? Double stars. You can find three attractive binocular pairs forming a neat row in the northeastern reaches of the constellation.

The easiest of the three to find is Delta (δ) Bootis. However, this pair is also the most difficult of the bunch to split. Delta is one of those cases where the separation between the two stars is generous enough, but the brightness difference between components is large. The stars are 104″ apart, but the 7.9-magnitude secondary is competing for attention next to its 3.6-magnitude, golden-yellow companion. Still, in my 10 × 50 binoculars, the secondary star was easy to see after I initially found it.

Moving a little more than one binocular field north-northeast from Delta, we come to the prettiest pair in the constellation, Mu (μ) Bootis. Mu's stars are separated by nearly the same amount as Delta's, yet in this case a 6.5-magnitude secondary star (itself, a telescopic double) holds its own easily next to a 4.3-magnitude primary. In my 10 × 50s, the brighter component shines with a slightly bluish tint.

The last stop on our Boötes tour is the wide, bright pair, Nu (ν) Bootis. These 5th-magnitude stars are more than 10′ apart, but what makes them appealing is their contrasting colors. I see the southern component (ν^1) as orange and its neighbor (ν^2) as a delicate blue — like a wider, subdued version of the famous colored double Albireo, in Cygnus.

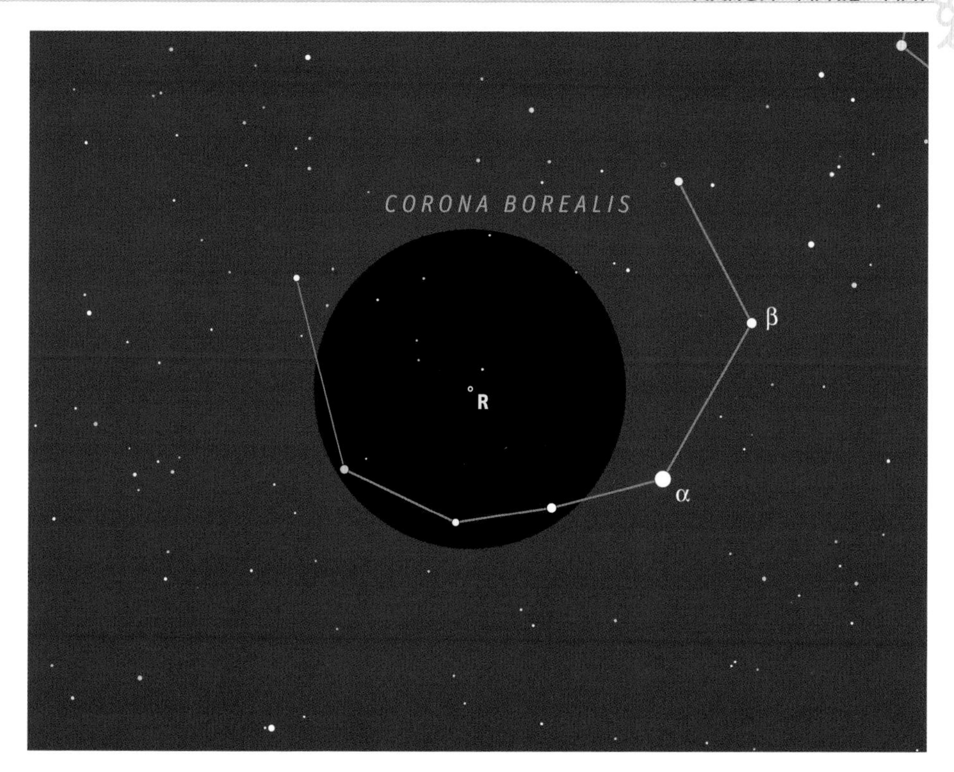

R Cor Bor's Vanishing Act

A peculiar thing happens from time to time to an otherwise unremarkable 6th-magnitude star in Corona Borealis. The star begins to fade gradually, almost imperceptibly at first, and then much faster. A month later the star is so faint that it can no longer be seen in binoculars. Then, just as quickly as it dropped out of sight, the star returns. Six weeks later it once again shines at the verge of naked-eye visibility as if nothing had ever happened. This is a typical unscheduled performance by the variable star R Coronae Borealis.

Lots of stars undergo periodic brightness changes, but among variable stars R CrB is an oddity. It will stay at magnitude 6.0 for months or even years, then suddenly fade to magnitude 14 or even fainter. These fade-outs are believed to result from carbon soot condensing in the star's atmosphere. After a few weeks or months, this obscuring material dissipates and the star returns to its normal brightness.

Following the ups and downs of this star takes only a moment. Why not regularly check in on R CrB at the start of your binocular observing sessions? You never know when it might vanish again.

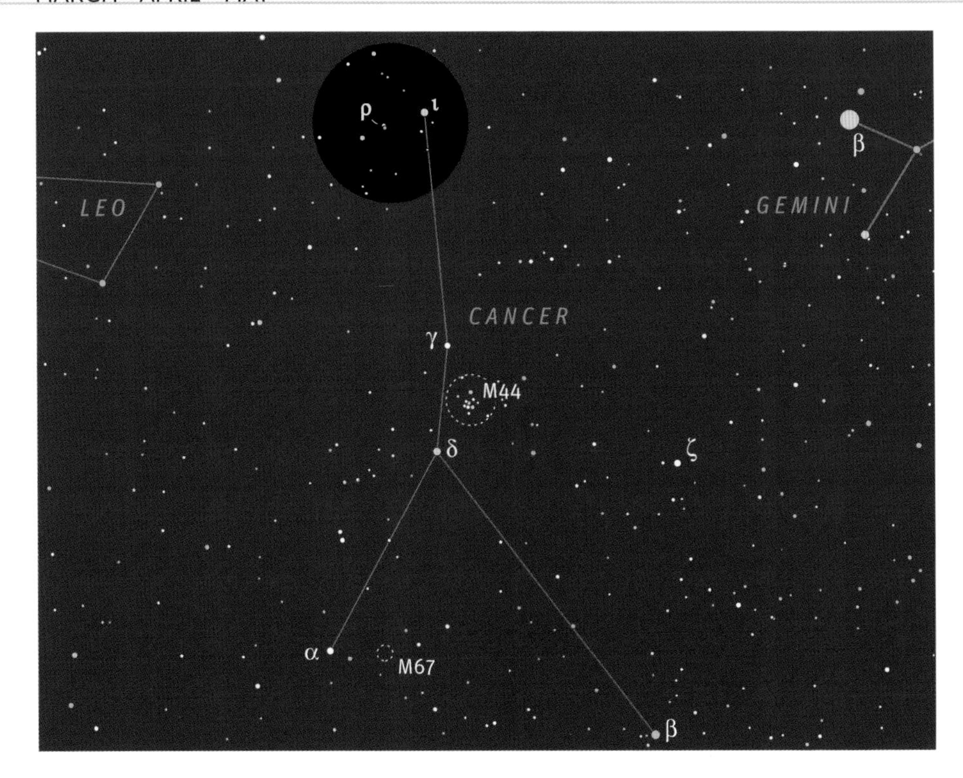

Two Doubles in Cancer

Even very high-quality binoculars can't match the magnification and resolution of a small telescope. However, this doesn't mean that if they are mounted or electronically stabilized, binoculars can't be used for viewing doubles — it just means that your targets are limited to wide, bright pairs. Fortunately, there are quite a few of these. Indeed, two lie within a single binocular field in the northern reaches of Cancer. One of the pairs is easy, the other difficult.

The easy double is the pair 55 and 53 Cancri, together known as Rho (ρ) Cancri. The star 55 Cancri was one of the first discovered to have an extrasolar planet, and it's now known to have four. Any binoculars will show this pairing of 6th-magnitude stars separated by a generous 278″ (arcseconds). The actual magnitudes of the components are 5.9 and 6.3. A brightness difference this slight is difficult to detect — especially at low magnifications. Can you tell which is the brighter of the two?

A little more than 1° to the west of Rho is the difficult double Iota (ι) Cancri. The difficulty arises from two factors. First, the stars in this double are only 30″ apart — a separation that is right at the resolution limit for 10× binoculars. Second, the primary star shines at magnitude 4.0, while its companion is only magnitude 6.5 — a 10-times difference in brightness. These two factors combine to make Iota a challenging pair to split. In my 10 × 30 stabilized binoculars I can see the companion star, but barely. Most of the time it looks like a slight optical flare coming off the brighter star, and I might easily overlook it. In my 15 × 45 stabilized binoculars, though, the secondary is not too difficult to separate from the primary's glare — the extra 5× really makes a difference.

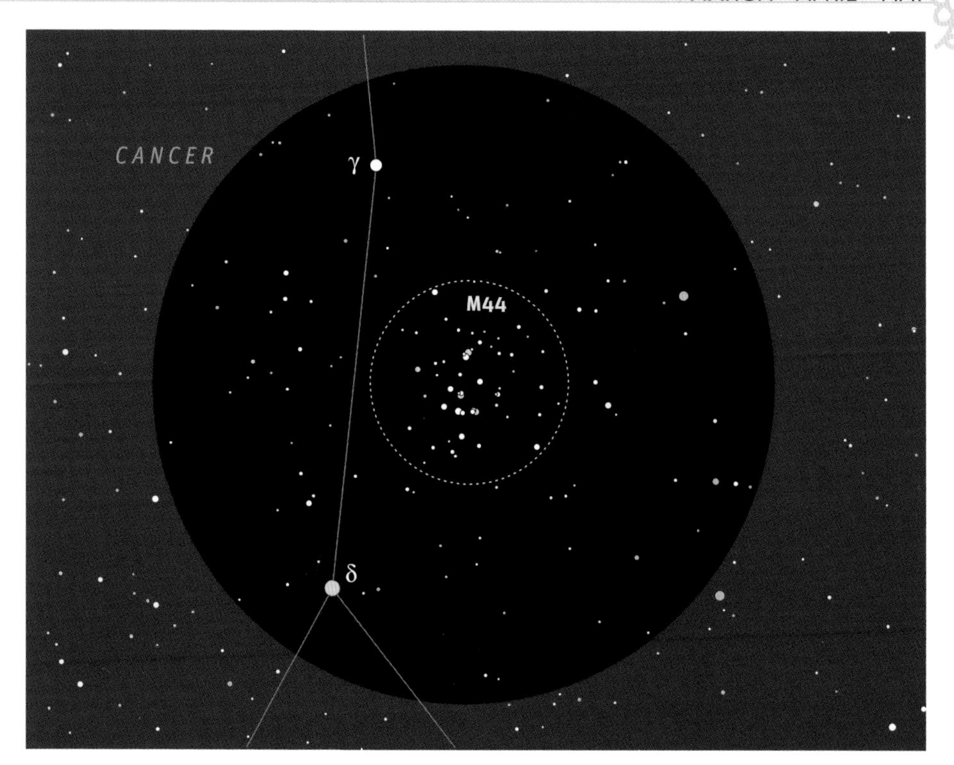

The Beehive Cluster

Although cancer is a dim constellation, it contains one of the northern sky's best binocular sights: M44, otherwise known as the Beehive Cluster. It's a particularly fine target for city-bound binocular observers. The cluster's 10 brightest stars shine at 7th magnitude or brighter and are sprinkled evenly across more than 1° of sky. With a little care, another dozen cluster stars can be made out with steadily held 10 × 50 binoculars. M44 is an impressive sight made all the more striking by its relatively barren surroundings. Unlike so many of the winter clusters now slipping into the west, the Beehive isn't set against a rich background of Milky Way stars.

Whenever I view the cluster I see a lop-sided box, looking like a miniature of the constellation Corvus, surrounded by a ring of stars. The overall appearance strikes me as less like a beehive than a celestial crab — the box is the body, and the ring stars mark the tips of the crustacean's legs and pincers. But it's all too easy to draw patterns in the sky. Perhaps my imagination is simply making a subconscious connection between the cluster and its constellation.

M44 is situated on the ecliptic, so it's frequently visited by the Moon and planets. When they pass by, the nightly motions of these solar-system visitors make for an exciting binocular show!

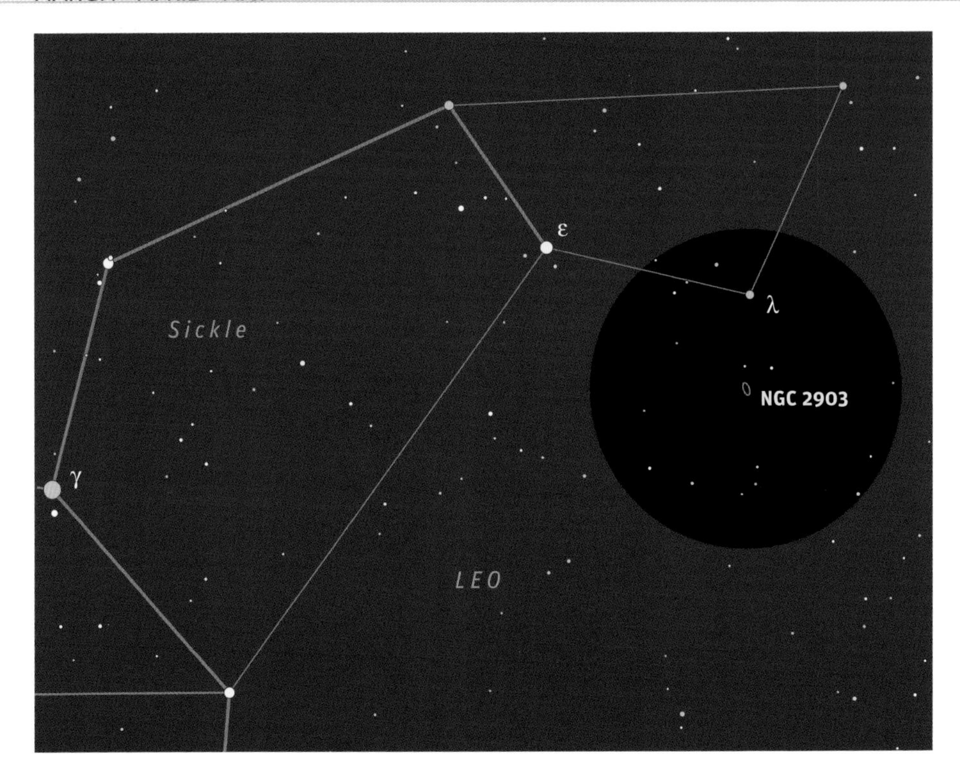

A Visit to NGC 2903

Whenever the Milky Way hugs the horizon, as it does late in the evening at this time of year, we get a wide-open view of the universe beyond the confines of our home galaxy. And what can we see? More galaxies. Indeed, a quick perusal of the NGC or Messier lists shows that most deep-sky objects are galaxies. Because they are far away, they are faint. As a result, only a handful of these distant island universes are interesting binocular objects. NGC 2903, near the Sickle of Leo, is one little-known example.

In 10 × 50 glasses, NGC 2903 appears as a tiny smudge of light 1½° south of 4th-magnitude Lambda (λ) Leonis. That doesn't sound impressive, but it is. Consider that with minimal optical aid you are able to look through a skin of foreground stars, across an expanse of more than 20 million light-years, and actually see the glow of billions of suns whose combined brilliance has been subdued to 9th magnitude by a distance beyond comprehension. This kind of realization is central to the experience of visual astronomy, whether practiced with the naked eye, binoculars, or a monstrously large telescope.

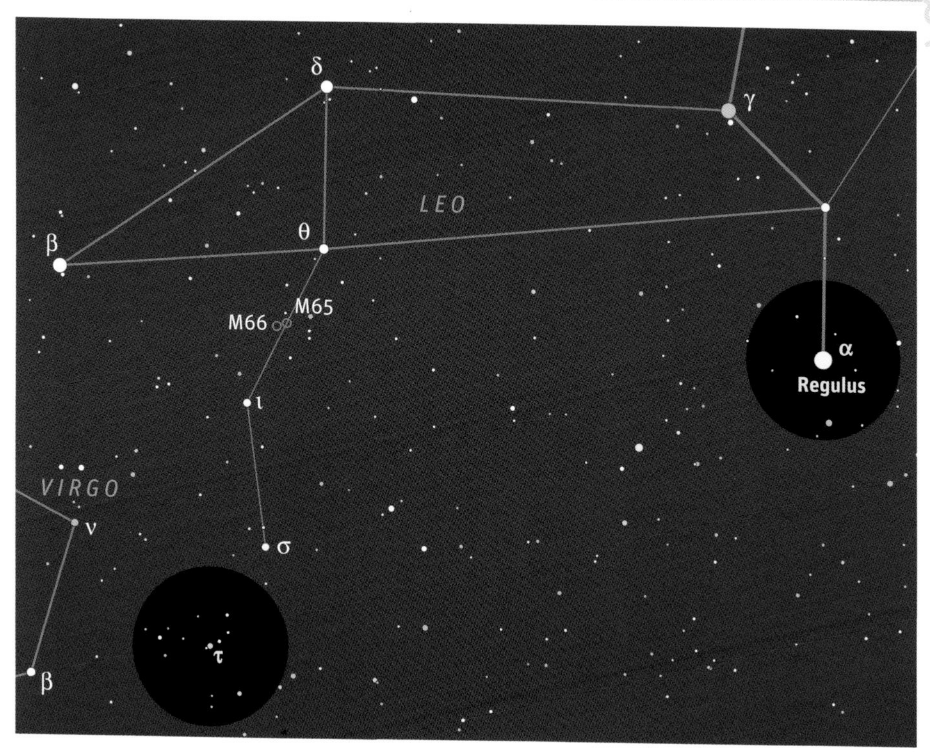

A Pair of Leo Doubles

When it comes to double stars suitable for ordinary binoculars, a couple of factors seriously limit the number of candidates. First off, both stellar partners have to be reasonably bright, and secondly they have to be far enough apart that they can be resolved at low power. Two doubles in Leo that make the grade are Regulus and Tau (τ) Leonis.

Like Polaris, Regulus (α Leonis) is a well-known star that many observers are surprised to learn is double. The component stars are far enough apart (176″) that they're easy to split, but the companion can be tough to see because it is an 8th-magnitude glow parked next to a 1st-magnitude beacon. Still, unless your skies are unusually bright, 10 × 50 binoculars will show the companion northwest of Regulus without too much difficulty.

Far easier to split is Tau Leonis. Although the pair are half as far apart as Regulus and its companion, you don't have to deal with wildly different brightnesses. Indeed, the hardest part is finding the double. The easiest route is to zigzag your way there one binocular field at time. Start at Theta (θ) Leonis, then head generally south to Iota (ι), then Sigma (σ), then finally to Tau.

The Tau pair consists of a 5.0-magnitude primary with a 7.5-magnitude companion lying a generous 89″ due south. This should prove an easy split in any binoculars. What makes this double even more appealing is that it is also part of an attractive little grouping of 6th- and 7th-magnitude stars.

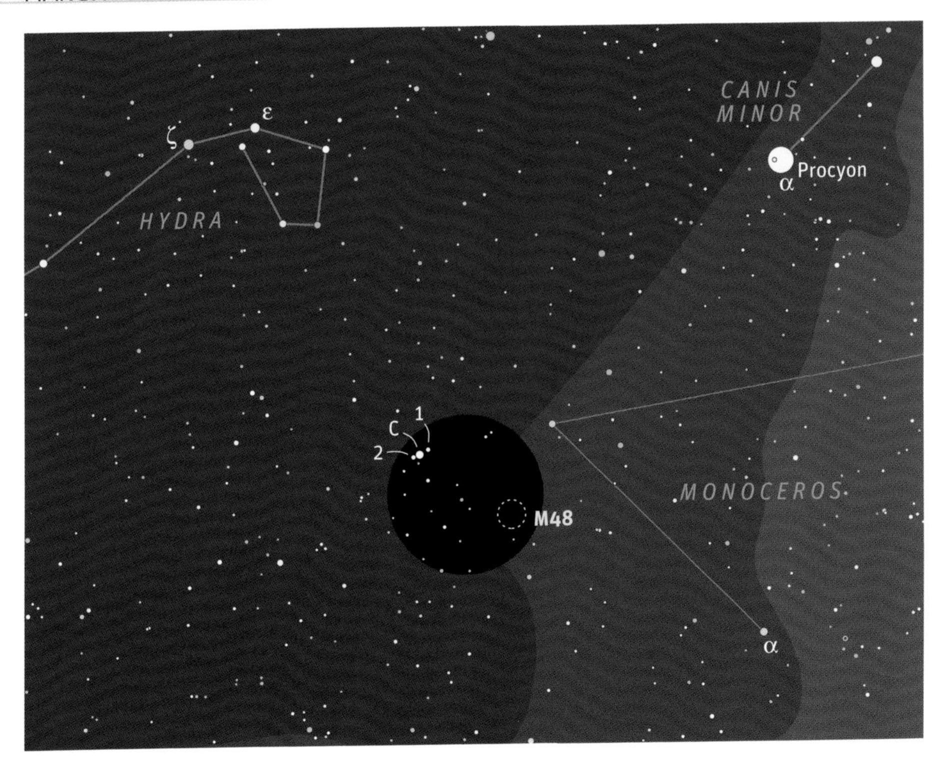

Star Cluster M48 in Hydra

With all the dazzlingly bright stars lying to the west, it's little wonder that the region between Hydra and Monoceros is often overlooked and that the open cluster M48, Hydra's most interesting binocular target, is not better known. In fact, the patch of sky surrounding M48 is so barren that deep-sky cataloger Charles Messier himself misplaced the cluster in his records by 5°.

The easiest way to find M48 is to first locate the odd little trio of 1, 2, and C Hydrae; the cluster is 3° southwest. Under bright suburban skies, M48 is a tricky find even when you have located the field. In my 10 × 50 binoculars I had to look carefully to make the cluster out against the light pollution. M48 appeared as a round glow from which one or two stars occasionally emerged.

A bright sky is exactly the situation in which more magnification really helps. Indeed, the view in my 15 × 45 image-stabilized binoculars was dramatically better. The soft haze of M48 was transformed into a pretty cluster in which nearly a dozen individual stars could be seen. If your skies are darker than mine, this is more like the view you can expect with 10× binoculars.

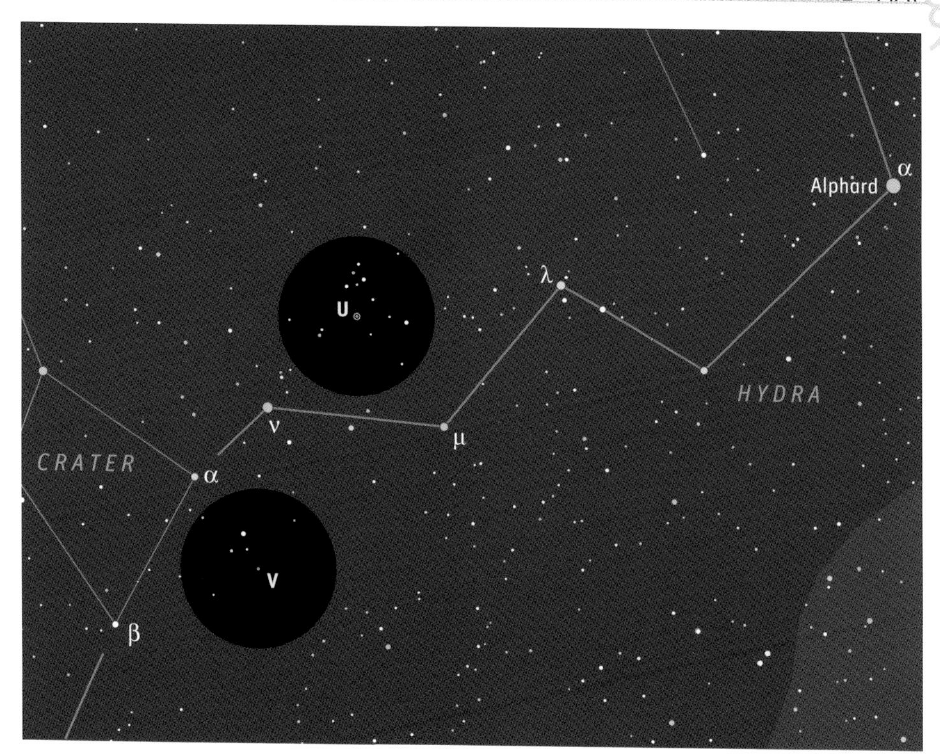

U and V Hydrae

Carbon stars are the most strikingly colorful stars in the sky. They are red giants, like Betelgeuse and Antares, but appear even redder due to the relative abundance of carbon in their atmospheres. The carbon-rich molecules act as a red filter, blocking the shorter (blue) wavelengths of a star's light.

One of the brightest carbon stars lies south of Leo in Hydra. Using the map above, start at Alphard, Alpha (α) Hydrae, and carefully trace your way eastward along Hydra's winding form until you are aimed midway along the line from Mu (μ) to Nu (ν) Hydrae. Look 3° north of this point. There you should see U Hydrae, a 5th- or 6th-magnitude gem glowing fire orange next to a lovely, curving row of 6th- and 7th-magnitude stars. Try defocusing your binoculars slightly. Spreading out the light of a star usually makes its dominant color more apparent.

About 8° to the south-southeast lies another carbon star, V Hydrae. It has been ranging from magnitude 6 to 10 in recent years with a period of about 550 days. To my eyes, V Hydrae is much redder than its brighter neighbor. What do you see?

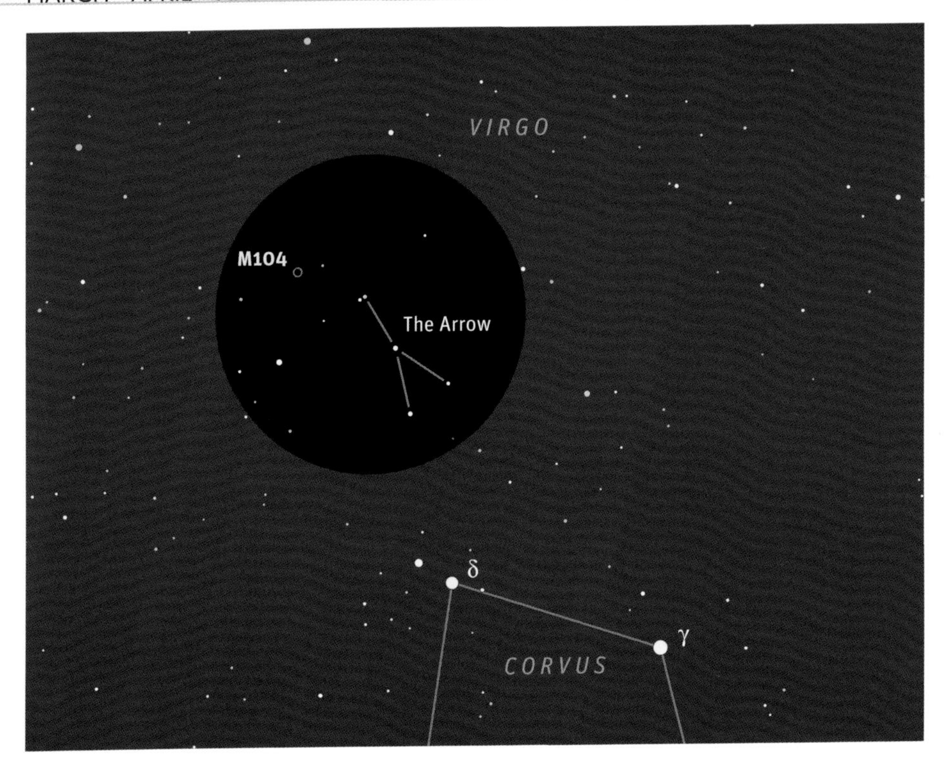

M104 and the Great Beyond

"How far can you see with that thing?" Anyone who has used a telescope to show the night sky to the public has doubtless heard that question more than once. There's something about a telescope that fires the imagination, but even modest binoculars can feed the desire to "see far."

So how far *can* you see with binoculars? Unfortunately, there's no simple answer. But given the fainter-is-farther nature of the universe, it is pretty safe to say that the darker your skies and the better your binoculars, the farther out you will be able to see. The Andromeda Galaxy, M31, is a snap at 2.5 million light-years (see page 88). In general, the most-distant binocular objects are galaxies that lie well beyond our Local Group, including this month's highlight, M104, the Sombrero Galaxy in Virgo.

Although M104 is one of the brightest Messier galaxies (magnitude 8), it's a challenge to see in typical suburban skies. In my 10 × 50 binoculars, the galaxy's hundreds of billions of suns amount to little more than a tiny, slightly fuzzy blip of light. To locate it, the easiest route is to start from Gamma (γ) Corvi and follow a string of 7th-magnitude stars 5° northeast, where you will find a little arrow-shaped asterism that points just a bit west of the galaxy. (Be careful not to confuse a nearby pair of faint stars for M104, though.)

Even if it isn't visually stunning in binoculars, by viewing M104 you are looking some 28 million light-years out into the cosmos. Keep that in mind the next time someone asks how far you can see with "that thing."

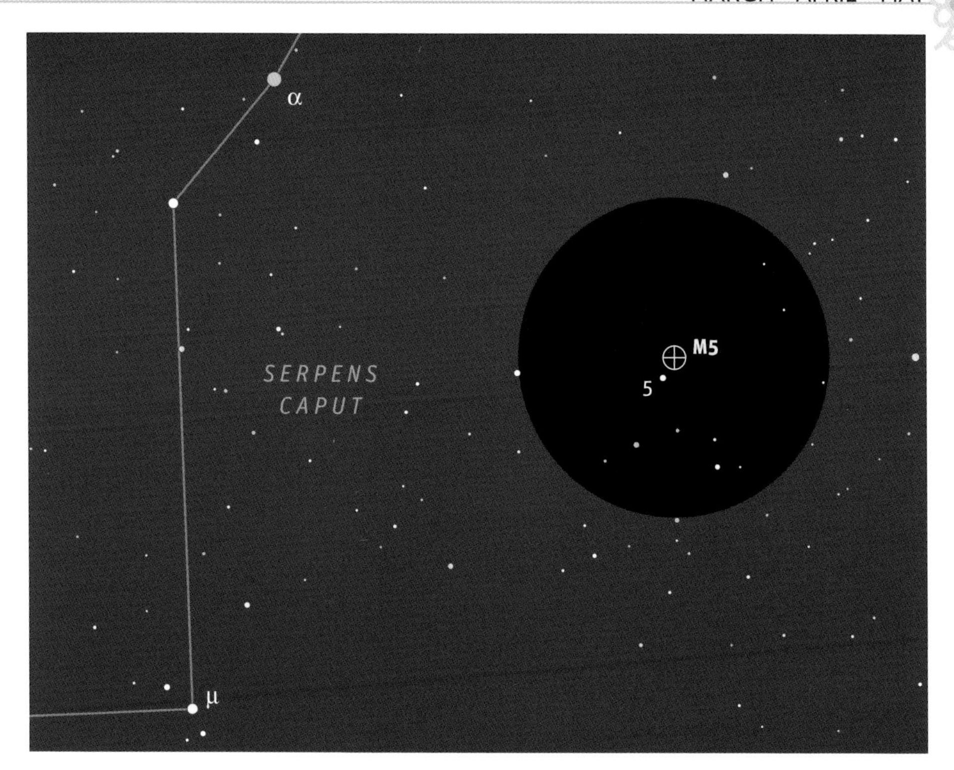

Ancient Globular M5

Bright globular clusters are among the most spectacular telescopic sights. Unfortunately, with few exceptions, they really don't come across very well in ordinary binoculars. The problem is resolution. Binoculars, with their small objective lenses and low magnifications, simply don't have enough horsepower to bust open these tightly packed star cities. That said, you can still enjoy the thrill of the hunt and the pleasure of quiet contemplation that accompanies the view.

One of the sky's finest globular clusters is M5, located in Serpens Caput. Although it isn't close to any conspicuous naked-eye landmarks, the cluster is bright enough to be swept up by scanning the general vicinity. Indeed, at magnitude 5.7, it is the brightest globular in the Northern Hemisphere and the third-brightest Messier globular — only M22 in Sagittarius (page 77) and M4 in Scorpius (page 74) outshine it. I usually locate the field by imagining the cluster as completing an equilateral triangle with Alpha (α) and Mu (μ) Serpentis.

Since it's situated right next to 5th-magnitude 5 Serpentis, M5's nonstellar nature is obvious in my 10 × 50 binoculars. The cluster appears as a conspicuous little blob of light with a bright, almost starlike nucleus. That might not sound impressive, but keep in mind that when you gaze at this globular you are looking at an unspeakably ancient object — a vast collection of stars likely more than twice as old as our Earth.

Akira Fujii

3

JUNE • JULY • AUGUST

PLANETARY NEBULA
GLOBULAR CLUSTER
DIFFUSE NEBULA
OPEN CLUSTER
VARIABLE STAR
GALAXY

ABOUT THE CHARTS:

Each of the star maps in this chapter has been rendered at one of three different scales: the wide-field charts to magnitude 7.5, the medium-scale charts to magnitude 8.0, and the close-up charts to magnitude 8.5. Regardless, the darkened circular area always represents the field of view for typical 10 x 50 binoculars.

Resolving Nu Draconis

Nearly overhead is the lovely double star Nu (ν) Draconis, the faintest of the four stars making up the head of Draco, the Dragon. For binocular viewing, this is one of the sky's prettiest pairs — twin white suns separated by only a bit of black sky. How distinctly you will be able to see both stars will depend on the resolving power of your binoculars — and on how steadily you can hold them.

For telescopes, resolving power is often described using the Dawes limit. For 50-millimeter objective lenses, the Dawes limit says that, in theory, we should just be able to split a double star separated by only 2.3″. Nu Draconis's components are separated by 62″ — which ought to be very easy! But the Dawes limit applies only when you use very high magnification, far beyond that of binoculars. Take a close look at Nu and you will see that this is indeed a tougher split than the Dawes limit would suggest.

A better guideline for binocular resolving power is to divide 300 by the magnification. For example, with 7× glasses, you should be able to just split a pair separated by 43″. This indicates that Nu Draconis will be splittable, though close. For a 10× pair Nu Draconis will be much easier, since binoculars that magnify by this amount should be able to resolve equal-brightness double stars separated by only 30″.

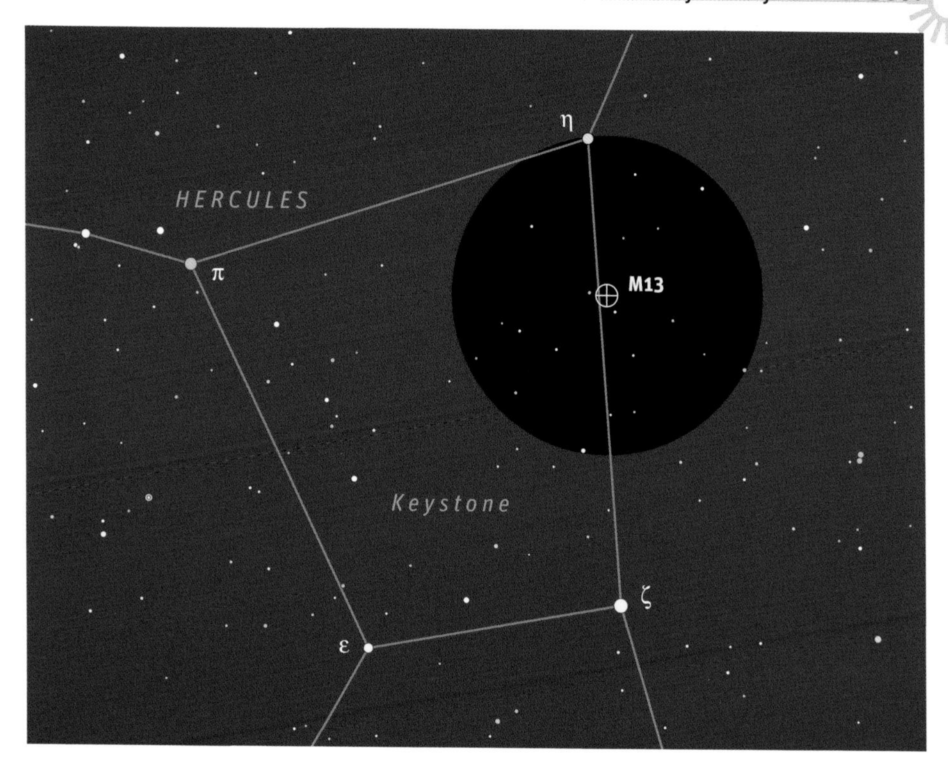

The Great Hercules Cluster

Of the globular star clusters lying north of the celestial equator, M13 in Hercules is one of the finest — if not *the* finest. While many other globulars are visible from midnorthern latitudes, some arguably even more interesting in binoculars, only a couple are as favorably positioned as the Great Hercules Cluster. In early summer this ball of a half million stars passes nearly overhead. For binocular viewing this is a mixed blessing. On the one hand, the cluster is unlikely to be obstructed by houses and trees. But you will probably have to lie on the ground or use a reclining chair when viewing it to keep the experience from becoming a literal pain in the neck.

Finding M13 is easy. Simply aim your binoculars about a third of the way along a line extending from Eta (η) to Zeta (ζ) Herculis, the western stars of the Keystone asterism in Hercules. Under dark skies, the cluster is very dimly visible to the unaided eye, but even suburbanites should have little trouble finding M13 with the smallest binoculars. The trick is to actually identify it. The giveaway is that the cluster looks slightly fuzzy — like a 6th-magnitude star that refuses to come to a sharp focus. This fuzziness becomes increasingly obvious with greater magnifications. Even with 7× binoculars M13 is conspicuously nonstellar, and 15× binoculars begin to show hints of the object's telescopic splendor.

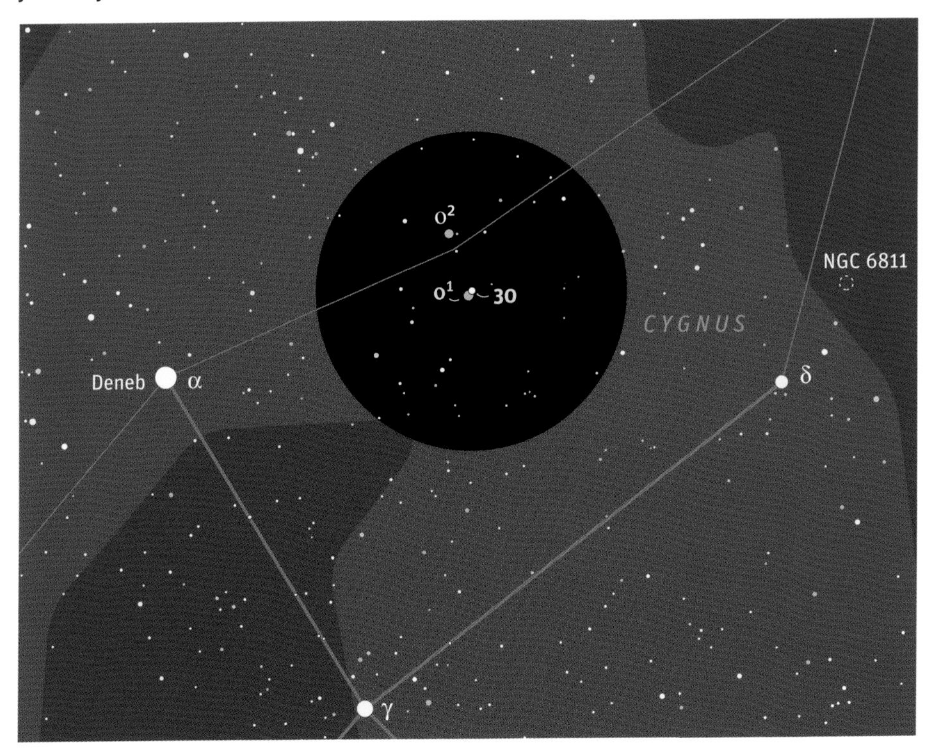

A Colorful Cygnus Triple

In the July 2001 issue of *Sky & Telescope*, Alan Adler listed 50 of the season's prettiest double stars and their optimum viewing magnifications. Although binoculars and double-star observing are not often thought of together, Adler's list shows that some of the sky's best pairs are suitable binocular targets. Indeed, seven of his selected doubles have optimum viewing magnifications in the range of common binoculars. Among them is the triple gem Omicron^1 (o^1) Cygni, now high in the evening sky for midnorthern observers.

This triple consists of a golden 4th-magnitude star (Omicron^1, also called 31 Cygni) widely separated from its bluish white, 4.8-magnitude companion (30 Cygni). If you have trouble seeing their colors, try defocusing your binoculars slightly — this trick spreads out a star's light and makes subtle hues easier to see.

Almost lost in the glow of Omicron^1 is the system's third member — a 7.0-magnitude sun that displays no distinct color. Steadily held 10× binoculars will show all three, but 7× glasses will have a harder time separating the faintest star from Omicron^1.

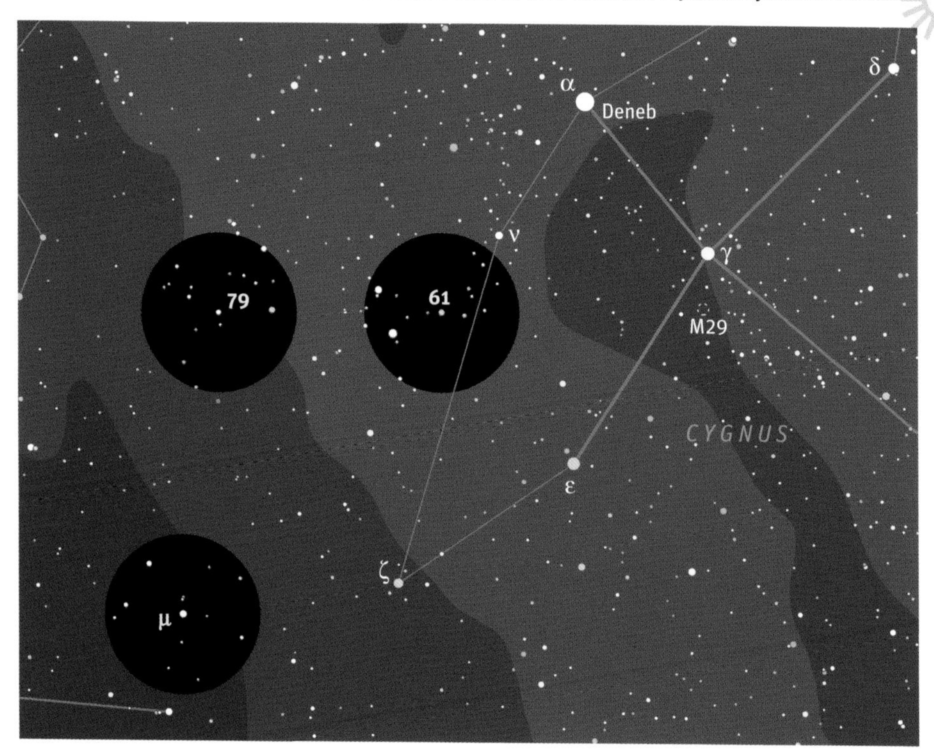

Three Cygnus Doubles

Tucked away in eastern Cygnus is a trio of binocular double stars that neatly demonstrate the challenge and the pleasure of observing stellar pairings. As a rule of thumb, the closest pair of equally bright stars you can split with steadily held binoculars can be estimated by dividing 300″ by your binoculars' magnification. For example, 10× binos should just be able to resolve stars separated by 30″ (300/10). But do actual observations bear this out?

Our first Cygnus double is Mu (μ) Cygni. Its stars (magnitudes 4.4 and 7.0) are separated by a generous 198″ and, as expected, they are an easy split in my 10 × 30 stabilized binoculars. The duo is set in an attractive field, with Mu forming a nearly equilateral triangle with two other 5th-magnitude stars lying a couple of degrees to the east. While you're there, look a half degree due north of Mu to see a slightly wider double consisting of twin 7th-magnitude stars.

About 9° north lies 79 Cygni. This pair of 5.7- and 7.0-magnitude stars are closer together (150″) yet not much more difficult to split than Mu. This is because there is only 1½-magnitude difference between the component stars compared to 2½ magnitudes for Mu.

The last double on our list is 61 Cygni — a pair of 5.2- and 6.0-magnitude stars separated by 31″. Our rule of thumb suggests that this will be a tough split for 10× binoculars, and indeed it is. I was able to separate the components only when my binoculars were perfectly focused. However, my 15 × 45 image-stabilized binoculars easily resolved the pair. And an attractive sight it is, too — stellar neighbors of the Sun side by side on a rich Milky Way background!

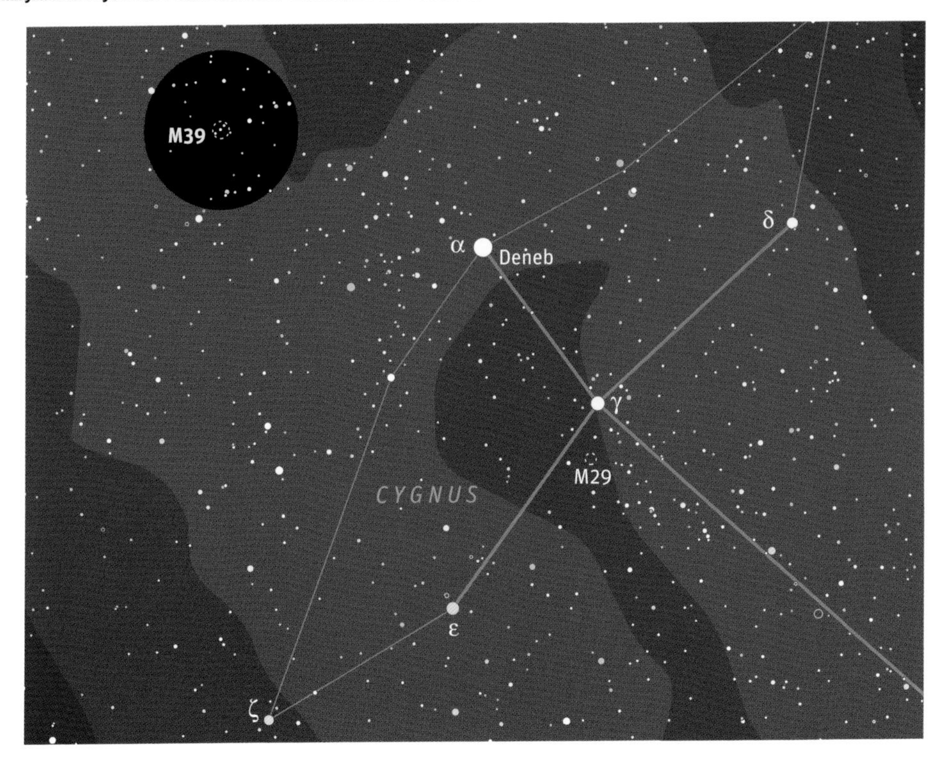

Open Cluster M39 in Cygnus

In spite of its size and prime location in the heart of the northern Milky Way, Cygnus is surprisingly devoid of binocular highlights. When viewed from a dark-sky location, the glowing stretch of star-shimmer running the constellation's length suggests all kinds of binocular riches that simply aren't there. Yes, there are some breathtakingly rich star fields, but only two Messier objects call Cygnus home: M29 and M39.

Of the two Messier open star clusters in Cygnus, M39 is definitely the more rewarding. From a reasonably dark location, my 10 × 30 image-stabilized binoculars show about a dozen stars with the brightest forming a distinct little equilateral triangle. The grouping is surprisingly easy to pick out from the rich starry background by sweeping the region along the Milky Way northeast of Cygnus's leading light, Deneb. Under suburban skies M39 loses a little of its charm, but it's still pretty easy to find. This is because the cluster boasts seven stars brighter than 8th magnitude.

It is perhaps ironic, but the darker your skies, the less likely you are to spend much time with M39. With the Cygnus Star Cloud beckoning nearby, it's tempting just to cruise the region and enjoy it for its own sake. But for suburban observers, M39 is the only game in town.

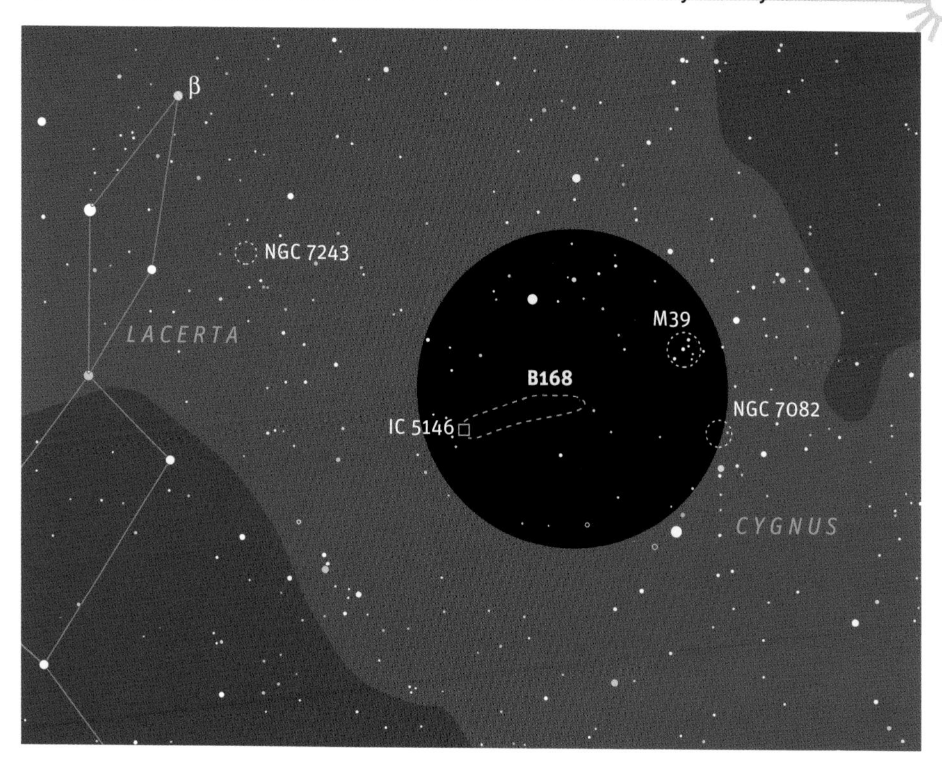

The Road to the Cocoon

I love dark nebulae. Seeing these clouds of interstellar dust silhouetted against a rich star field seems to give the Milky Way a three-dimensional appearance and make the galaxy look that much more *real*. One of my favorite dark nebulae is Barnard 168, or as I call it, the Road to the Cocoon Nebula. When I'm hunting for the position of the Cocoon (IC 5146) in my finderscope, I follow this narrow finger of darkness to the emission nebula's telescopic field. B168 stands out even better in my 10 × 50 binoculars, which pull in fainter stars than my telescope's 6 × 30 finderscope. Binoculars will not show the Cocoon itself — it's simply too faint.

Seeing B168 is a challenge that requires skies free from light pollution. The easiest way to find this dark nebula is to head east-northeast from Deneb about 7° (about one binocular field) to the pretty open cluster M39, and then move the cluster off to the western edge of your field of view. B168 should be visible on the eastern edge of the field. Of course, the more prominent the Milky Way appears from your location, the easier it will be to find the Road to the Cocoon.

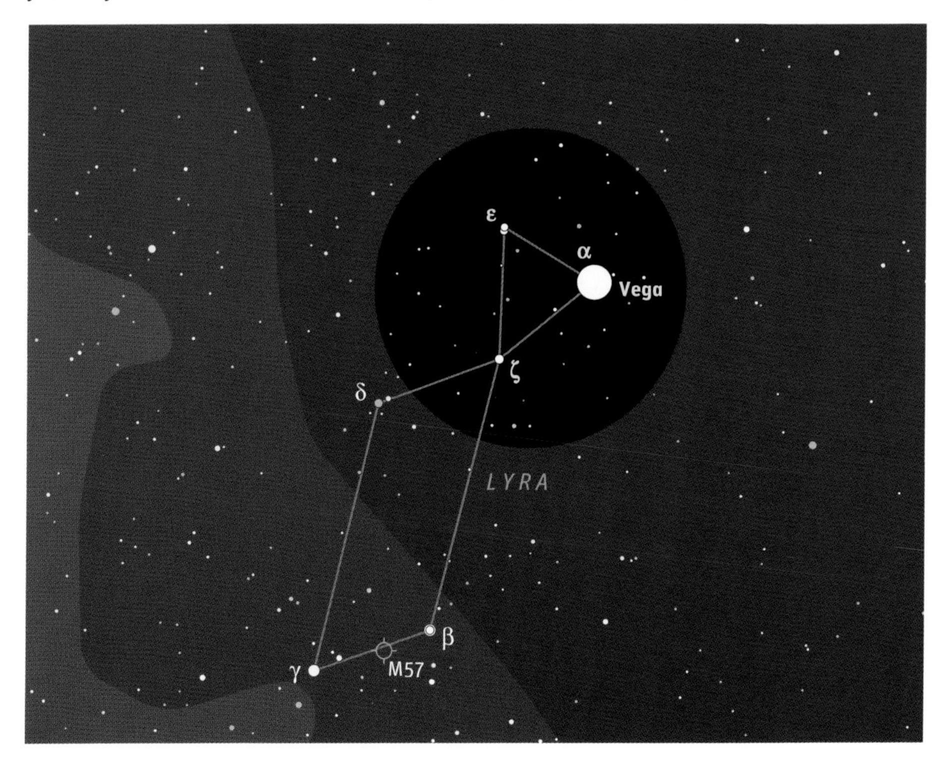

Summer's Other Triangle

Just about everyone has heard of the Summer Triangle, made of three of the season's most conspicuous stars: Vega, Deneb, and Altair. Summer's *other* triangle is a binocular grouping that also includes Vega, along with nearby Epsilon (ε) and Zeta (ζ) Lyrae.

The fifth-brightest star in the nighttime sky, Vega is dazzling even to the unaided eye — in binoculars it is doubly so. To the northeast is the famous Double-Double, Epsilon Lyrae. Any binoculars will easily show Epsilon^1 and Epsilon^2 as a wide pair closely matched in brightness and white color. However, to see each of these as double, you need the magnification of a telescope.

Due south from Epsilon is the third corner of our triangle, Zeta. It is another binocular double, but it's more difficult to split than Epsilon. Not only are the stars close together, but the main star is 3½ times brighter than its companion. Unequal pairs are much more difficult to separate than equal-brightness doubles. You'll need at least 10× binoculars mounted on a steady support and sharply focused to succeed.

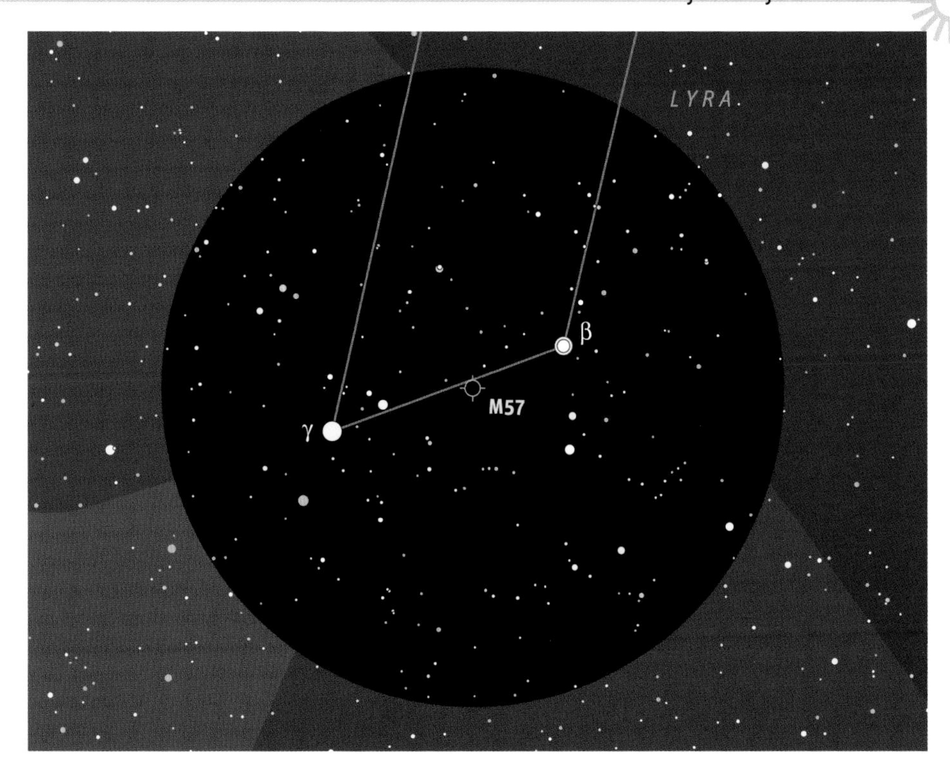

M57 and Expectations

Finding difficult deep-sky objects tests your observing skills, your optics, your skies, and sometimes your patience. When you can't see your target, the last of these really comes into play. During moments of frustration, though, it's worth pausing to take a deep breath and consider that when you come up empty, there are only three possible reasons: you're not looking in the right place, the object you are seeking is too faint for your equipment or sky conditions, or your expectations are out of sync with reality. The Ring Nebula, M57 in Lyra, is a notoriously difficult binocular object and provides a prime example of how easy it is to go wrong.

Finding this planetary nebula's location couldn't be easier. As the detailed chart above shows, it lies between a pair of 3rd-magnitude stars. At magnitude 8.8, M57 isn't exactly bright, but it should fall easily within the range of 50-millimeter binoculars. So if you're looking in the right place with equipment up to the task, what does it mean if you still can't see it? We're left only with reason #3: expectations.

Often, when we try to see a deep-sky object for the first time, we do so without really considering what it should look like. If you approach M57 with the memory of your last telescopic view or some color photograph, you'll probably miss it in binoculars. The Ring is only 80″ by 60″ across. So, what you should be hunting for is something that looks like a faint, slightly out-of-focus star. If you keep that image in mind while you carefully examine the area, you'll find M57. It's really a matter of expectations.

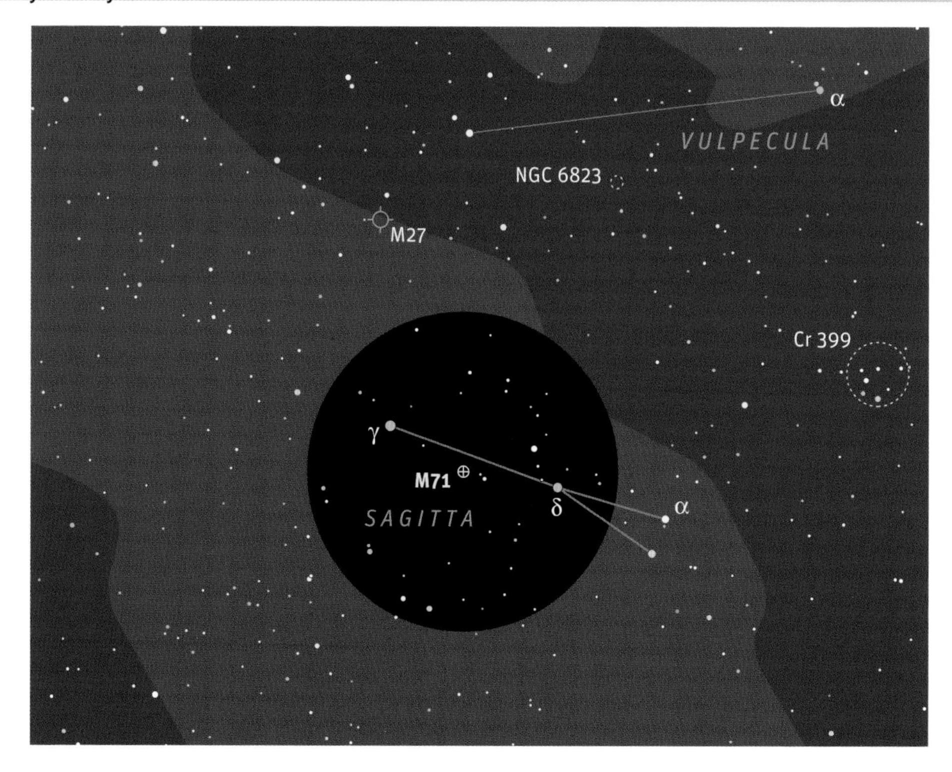

Splendid Sagitta

Is there such a thing as a binocular constellation? Yes — the diminutive grouping Sagitta, the Arrow. Both the unremarkable constellation Equuleus (the Little Horse) and Crux (the Southern Cross) take up fewer square degrees of celestial real estate, but of these Sagitta is the only one whose entire stick-figure shape (just 5° long) fits into the view of ordinary binoculars. And what a view it is! The little constellation's distinctive pattern is set against a rich Milky Way field sparkling with faint stars that's also home to a couple of binocular treasures.

Sagitta's connect-the-dots arrow consists of four stars of similar brightness, which makes the figure all the more conspicuous. In 7× or 10× binoculars the constellation alone forms a striking field, but there's more here than a quick glance will show. Look halfway between Gamma (γ) and Delta (δ) Sagittae (the stars that form the shaft of the arrow) and you should be able to spot the 8th-magnitude cluster M71. But is it a sparse globular cluster or an unusually rich open cluster? For many years astronomers were uncertain, but today there is little doubt that M71 is a relatively nearby globular lying about 13,000 light-years away.

Anyone scanning northwest of the Arrow's tail is sure to stumble across the Coathanger — Cr 399. Turn the page to read about this Milky Way surprise.

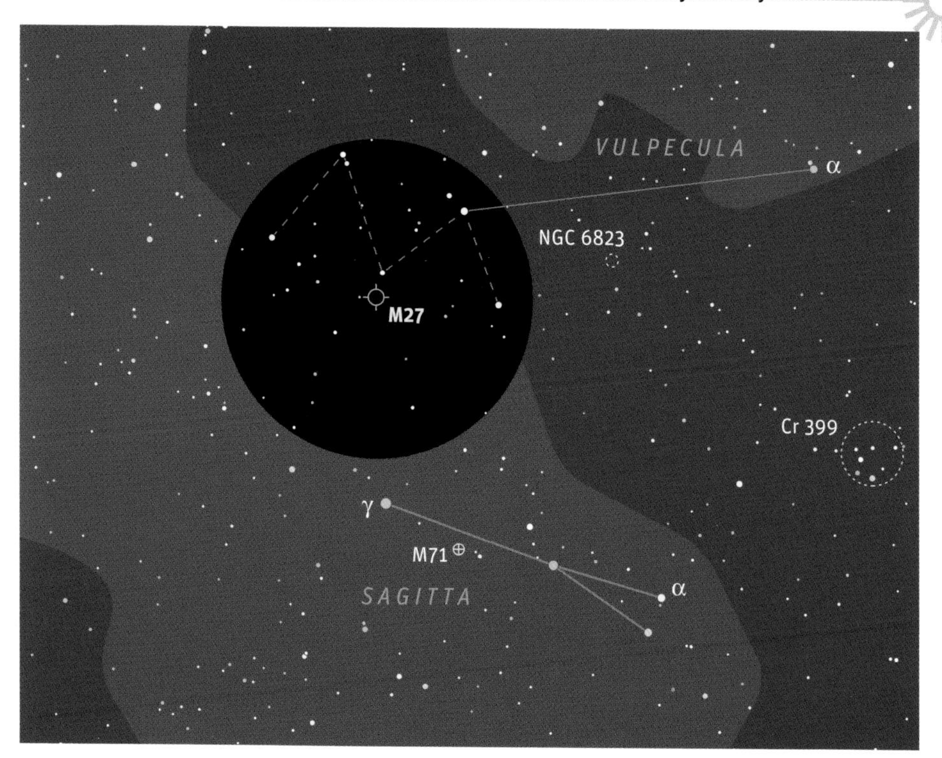

M27 in Vulpecula

On page 61 I described the binocular challenge of finding the Ring Nebula, M57, in Lyra. Its small apparent size means that you're unlikely to chance across it while casually sweeping the area. But the Ring is not unusual in this regard; the list of bright planetary nebulae big enough to appear nonstellar in binoculars is very short. In fact, there are only four planetary nebulae in the entire Messier catalog. Of these, M27 in Vulpecula (also known as the Dumbbell Nebula) is the biggest and brightest. It's 350″ across, indicating that it is relatively close — only 1,240 light-years distant.

The nebula glows at magnitude 7.3 in a rich swath of the northern Milky Way. This creates a problem in that Vulpecula is a small and indistinct constellation that doesn't exactly jump out at you. I have always found it easier to locate M27 by proceeding 3° due north from Gamma (γ) Sagittae, the star marking the tip of little Sagitta, the Arrow.

Once you have located the nebula's field, look for a small glowing disk situated under the middle star of a distinctive M-shaped asterism consisting of 5th-magnitude stars set against a myriad of fainter field stars. Binoculars magnifying 10× will easily show the nebula, and even 7× glasses should have little difficulty. However, ordinary binoculars do not have enough magnification to reveal the nebula's dumbbell shape.

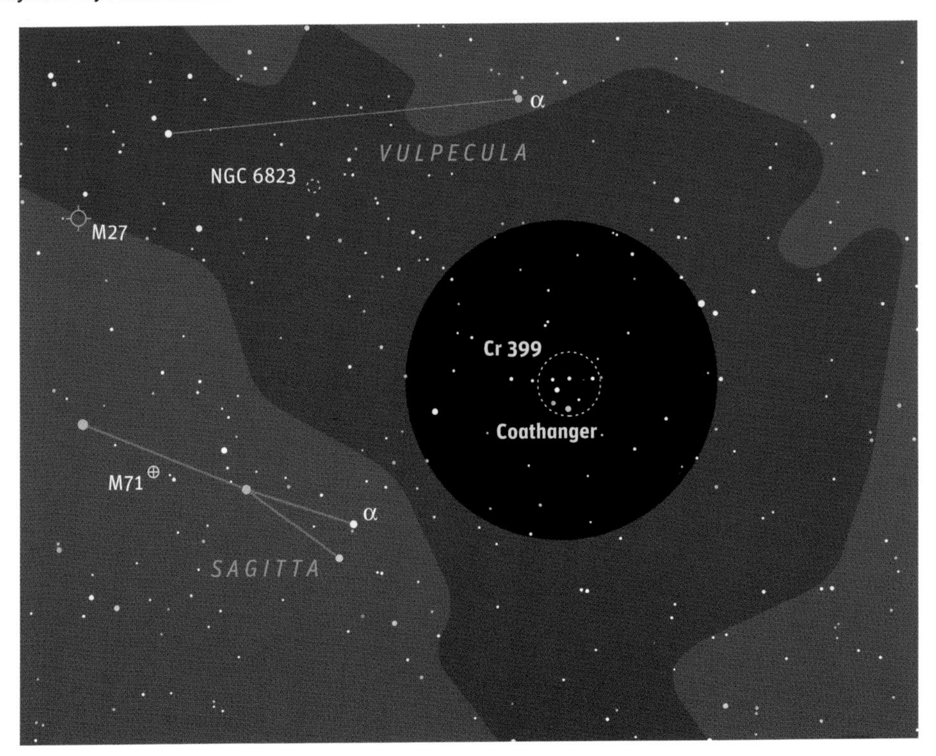

A Milky Way Surprise

The lingering warmth of summer evenings seems to engender a different kind of stargazing — slow, relaxed, contemplative binocular sweeps up and down the glowing length of the Milky Way. Sometimes these casual surveys turn up delightful surprises. Brocchi's Cluster (Cr 399) was such a find for me. I stumbled upon this little asterism many years ago as I scanned the region southeast of Cygnus with binoculars, while soaking up the ambiance of a country night from the comfort of a lounge chair. So conspicuous was this collection of stars that I could not believe I had never noticed it before. I was even more surprised to discover that it wasn't in my trusty *Norton's Star Atlas* and didn't even rate a Messier number!

Better known as the Coathanger (for reasons obvious upon first glance), the group contains about a dozen stars ranging from 5th to 9th magnitude. Despite appearances, the Coathanger is not a true cluster but a chance alignment of stars that are actually at very different distances. Nonetheless, it is a lovely sight easily within the grasp of any binocular. Look for it about 5° northwest of the distinct little constellation Sagitta, the Arrow, in the midst of the Milky Way.

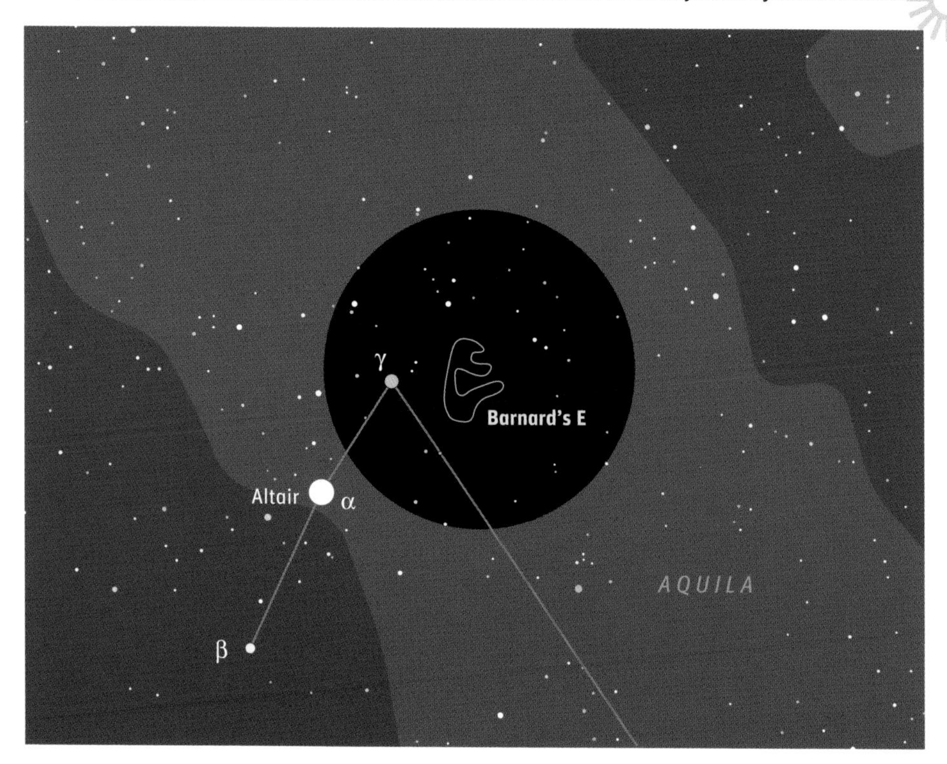

Barnard's "E"

To see this highlight you will need a sky as free from light pollution as you can find. Fortunately, at this time of year many families take to the road for camping trips to rural places where they can enjoy the Milky Way in its full splendor.

If you find yourself in such a location on a clear, moonless night, aim your binoculars nearly 3° (about a half-binocular field) to the northwest of Altair, the southernmost star of the Summer Triangle. Examine the region carefully. Silhouetted against the background glow of faint stars, you should see a small (about 1° high) E-shaped absence of stars — Barnard's E. The brighter the Milky Way from your viewing location, the more apparent this feature will be.

Barnard's E (also dubbed the "Triple Cave" nebula by Max Wolf, who discovered the object photographically in 1891) is a dark nebula — a cloud of interstellar dust and gas thick enough to block the light from stars behind it. Such opaque clouds were cataloged by Edward Emerson Barnard at the turn of the century. It seems fitting that one of the sky's most distinctive dark nebulae should bear his first and second initials!

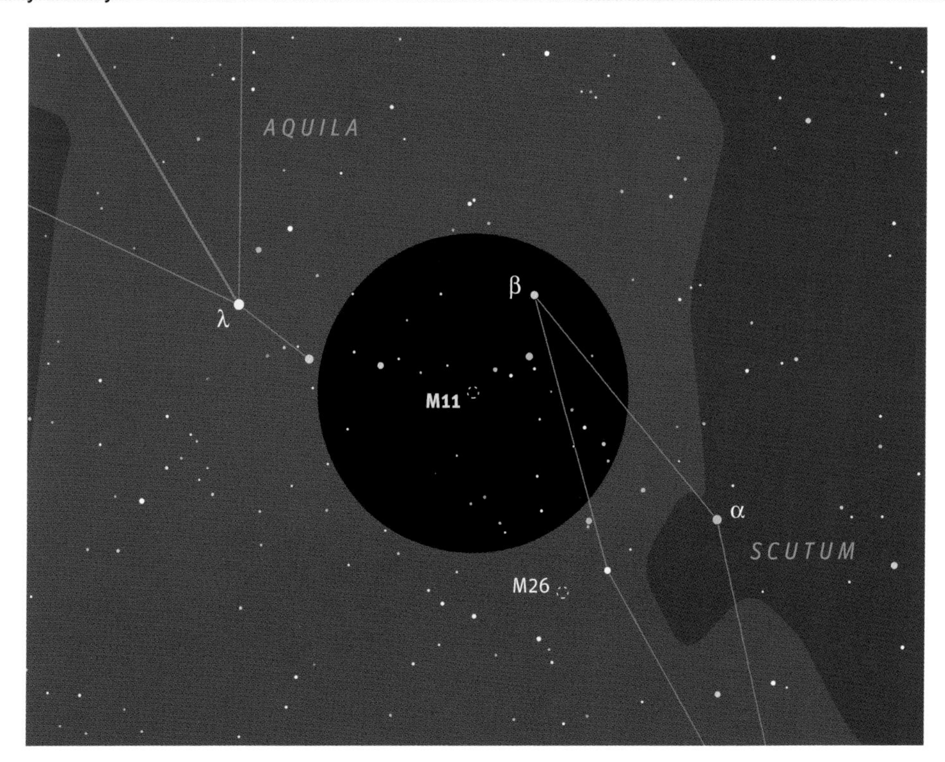

M11 in Scutum

Although the Messier catalog is known far and wide as a treasure-trove of deep-sky wonders, Charles Messier was actually a comet hunter. He began to compile a list of faux comets beginning in 1758 with the Crab Nebula, M1. In his small refractors (crude by today's standards), many of the objects he cataloged must have looked very cometlike — perhaps none more so than the rich open star cluster M11, also known as the Wild Duck Cluster.

Although M11 is in the small constellation Scutum, I find it easiest to locate by following the curving row of stars that make up the tail of Aquila, the Eagle.

No matter what your sky conditions, M11 should be a pretty easy catch. I have no trouble spotting it in 10 × 30 binoculars from my light-polluted backyard. And it really does look like a tailless comet! This illusion is enhanced by a relatively bright star slightly southeast of the cluster's center. In binoculars, M11 looks for all the world like a comet nucleus surrounded by a hazy coma.

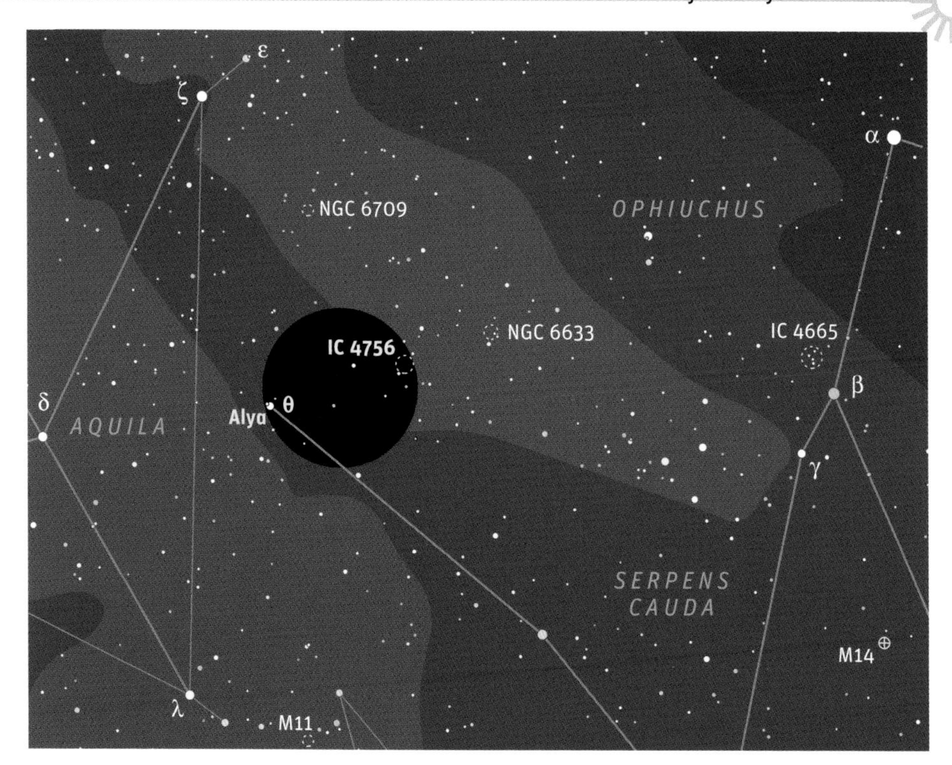

IC 4756 in Serpens

When we think of the Milky Way at this time of year, the constellations Sagittarius, Cygnus, Scorpius, and Aquila spring to mind most readily. Seldom do we consider the riches found in Serpens Cauda, the tail of the serpent situated to the east of Ophiuchus.

My favorite sight in this area is the big, splashy open cluster IC 4756. If you're randomly sweeping the region with binoculars under dark skies, you're likely to stop in your tracks when you come across this object. It appears as a glowing mass of faint stars — like a detached chunk of the Milky Way. As with so many clusters that lack prominent bright stars, city skies will make IC 4756 a more difficult find.

While in the neighborhood, check out the pretty double star Alya, Theta (θ) Serpentis, about 5° east-southeast of IC 4756. Alya is a striking, nearly equal-brightness pair of 4.5- and 5.4-magnitude stars separated by only 22″ — image-stabilized or firmly mounted 10× binoculars are needed to cleanly split this double.

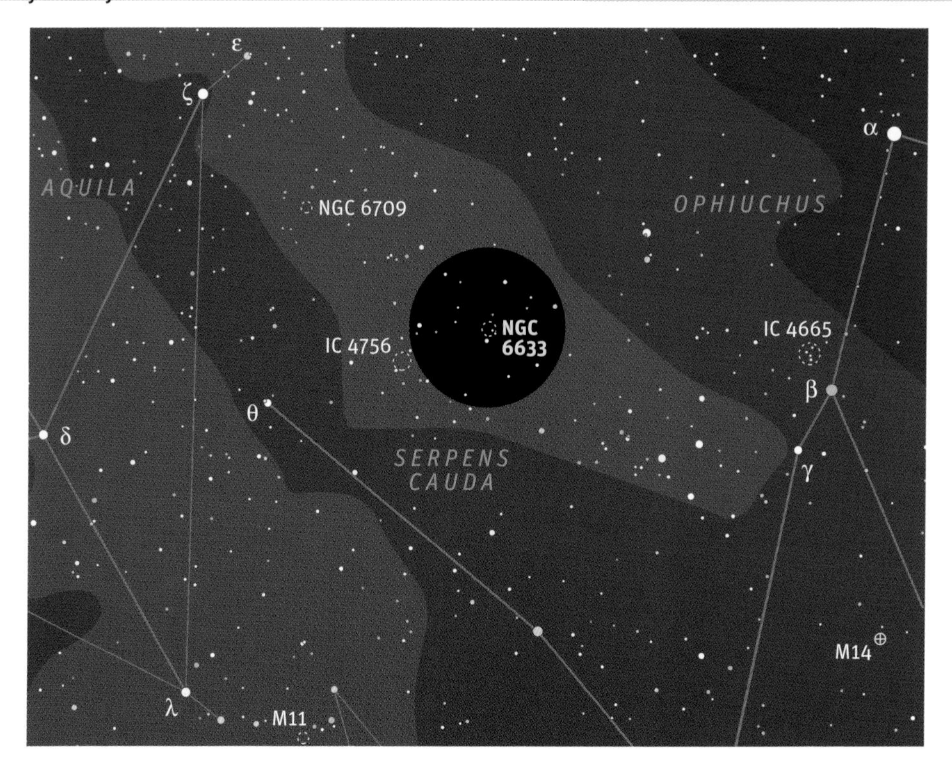

Three Star Clusters in Ophiuchus

Many binocular stargazers assume that the Messier list represents the best and brightest deep-sky objects the sky has to offer. But there are many worthy targets that lack Messier numbers. For example, perched off the eastern shoulder of Ophiuchus, not far from Beta (β) Ophiuchi, is a trio of fine binocular open clusters that were not cataloged by Charles Messier: IC 4665, IC 4756, and NGC 6633. They are certainly bright enough to have been visible in Messier's telescopes, but perhaps he, like many present-day observers, neglected this side road of the Milky Way in favor of the rich fields lying to the east in Aquila and to the south in Sagittarius.

Of the three clusters, NGC 6633 is perhaps the least conspicuous, but in my opinion it is the prettiest and well worth searching out. Sweep for it about 10° due east of Beta Ophiuchi. There, keen-eyed observers using binoculars with a magnification of 10× or more will be able to discern several stellar pairings forming a ragged ladder of stars aligned roughly northeast to southwest. The cluster's suns shine with a combined light equivalent to that of a single 4.6-magnitude star, which means that under a dark sky NGC 6633 should be faintly visible even without binoculars.

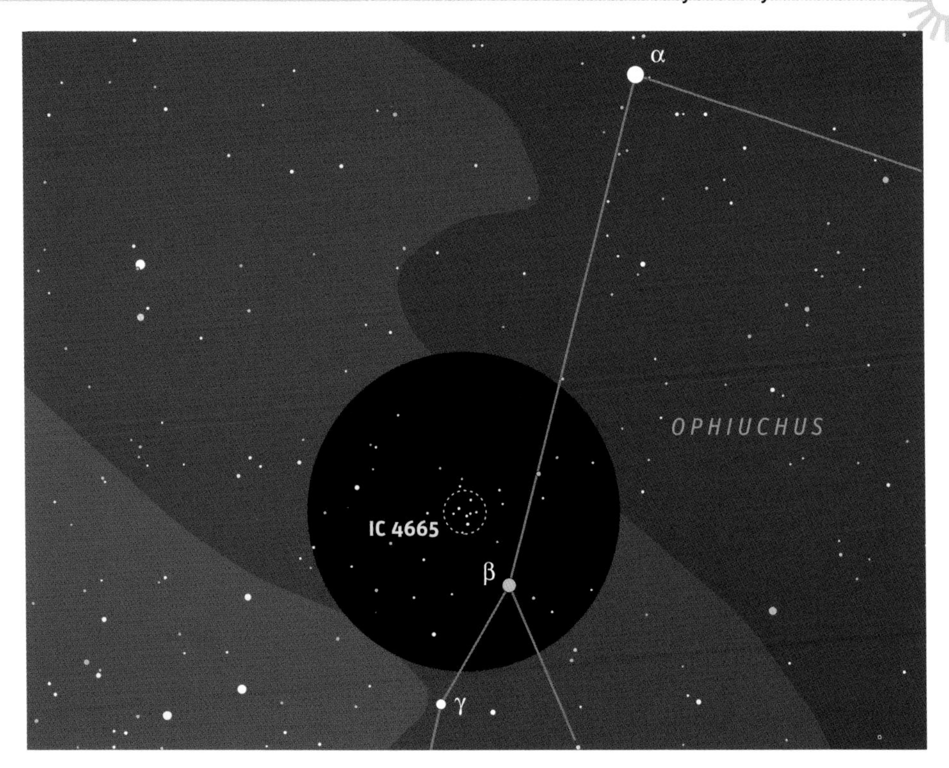

IC 4665 in Ophiuchus

Summer is a terrific time of year for spending moonless evenings under the Milky Way. The home galaxy's entire luminous expanse is dotted with all kinds of binocular prey — globular clusters, nebulae, and perhaps most rewarding, bright open clusters. Some are big and splashy, like the Double Cluster in Perseus (page 17); others appear as modest enhancements of the background star field. IC 4665 lies somewhere in between.

This cluster has the virtue of being easy to find. As indicated in the chart above, it's only a little more than 1° north-northeast of 3rd-magnitude Beta (β) Ophiuchi. The cluster is a nice binocular sight consisting of about a dozen 7th- and 8th-magnitude stars.

One of the interesting features of open clusters is how their stars form distinctive patterns and shapes. Stephen James O'Meara, in his book *The Messier Objects,* even visualizes "whimsical creatures" delineated by cluster stars. To me, IC 4665's brightest stars arrange themselves into a pair of boxes and a curving row that bends southward. What do you see?

A Pair of Ophiuchus Globulars

This season the Milky Way arches across the sky accompanied by its retinue of globular clusters. Most of these stellar swarms congregate around our galaxy's center, which lies in the direction of Sagittarius. Seven Messier globular clusters are found within that constellation alone. Surprisingly, the large (but otherwise unremarkable) constellation Ophiuchus contains as many Messier globulars as Sagittarius. How many of the Ophiuchus globulars you will be able to see depends on two things: the quality of your sky and the magnification of your binoculars. No question about it — higher magnification and darker skies will be a big plus as you try to pluck these clusters from the starry background.

Start your globular hunt by tracking down M10 and M12. These two Messier globulars have the largest apparent sizes of the Ophiuchus globulars and, as such, will be the ones that look the least starlike in binoculars. M10 and M12 are separated by only a little more than 3°, which means they will both fit comfortably within the field of view of most binoculars. Together, they make an attractive pair set against a rich starry background.

The proximity of M10 to M12 invites comparisons. What do you see? Are both the same size? The same shape? Although they are listed as having essentially the same magnitude (6.6 and 6.7, respectively) and apparent size, do they appear equally conspicuous to you? Addressing such questions is the difference between just *looking* and really *seeing* deep-sky objects. I don't want to prejudice you by offering my observations, but suffice it to say that the numbers rarely tell the whole story.

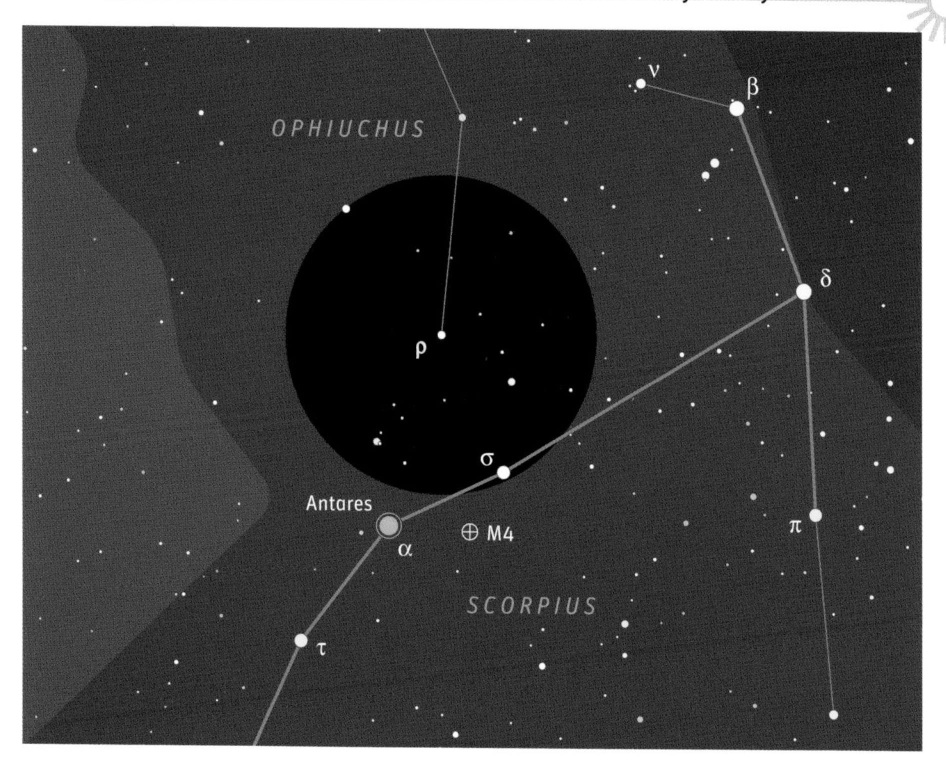

Rho Ophiuchi

If the star Rho (ρ) Ophiuchi seems familiar to you, it's probably because of the cloud of nebulosity that surrounds it. Indeed, one of the most spectacular and often-reproduced space images is David Malin's famous Anglo-Australian Observatory photograph of the Rho Ophiuchi region. Sadly, this colorful nebulosity is far beyond the reach of binoculars. But that doesn't mean there is nothing here for binocular stargazers — Rho itself is a pretty triple star.

It's easy to miss Rho in this rich swath of sky lying on the western fringe of the Milky Way, but if you place Antares and Sigma (σ) Scorpii at the bottom of your binocular field of view, Rho should be easy to spot just above the center of the field.

Seeing all three stars of this triple should be quite easy even in ordinary 7× binoculars. All are bright enough and far enough apart to be seen without difficulty. Rho's brightest sun is magnitude 5; its companions are each magnitude 7 and are separated from the main star by about 2½′ (arcminutes). In fact, with brilliant Antares blazing away and the globular cluster M4 glowing dimly nearby, all in the same field of view as Rho, avoiding distractions might be the hardest thing about observing this triple star.

A Solar Twin

We often read that our Sun is an "average" star, suggesting that the Milky Way is populated with plenty of other stars just like it. In fact, stars that closely match the Sun's brightness, color, size, age, and composition — virtual solar twins — are rather rare.

The nearby star that is most like the Sun is 18 Scorpii, according to a study by Laurence E. DeWarf (Villanova University) and five colleagues. It's about 18° north-northwest of ruddy Antares.

Imagine you are 46 light-years from home and looking back at the solar system. The Sun would be dimly visible to the naked eye and an easy binocular object. You check for color, perhaps expecting to see a pale-yellow tint, and are surprised to see only pure white — exactly how 18 Scorpii appears.

In spite of its similarities with our Sun, 18 Scorpii wouldn't make a perfect replacement. It is about 6 percent more luminous than Sol. A difference even this small would wreak havoc with Earth's climate.

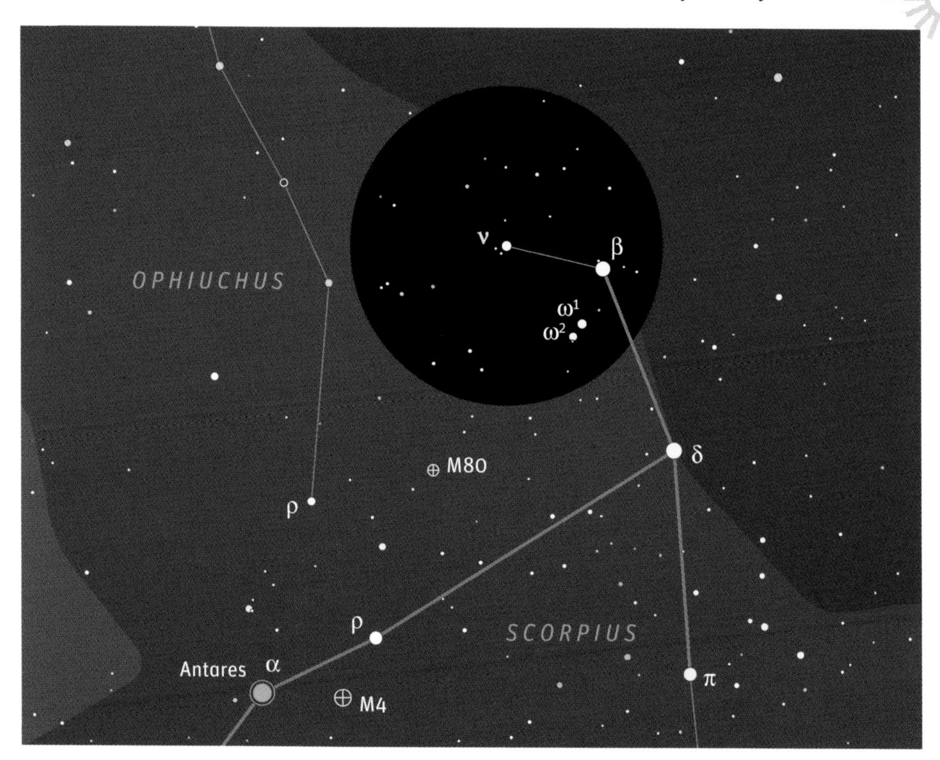

A Double for Steady Hands

On page 12 of this book, a number of simple yet effective binocular mounts are described. Although these devices rob some of the instant-observing appeal from binoculars, they handsomely repay that loss with improved views that reveal greater detail. The value of a binocular support is made plain by observing close double stars, like Nu (ν) Scorpii — one of three double stars that form an attractive triangle with Beta (β) and the wide double Omega (ω) Scorpii.

Nu's component stars are separated by 41″, which should make them a challenge for 7× binoculars but not too difficult for 10× glasses. The fact that the component stars differ in brightness by nearly 2 magnitudes increases the level of difficulty a notch or two. Try viewing Nu with hand-held binoculars. With 10× glasses you may glimpse its double nature, but only fleetingly. Now try steadying the view by leaning against a wall or propping your binoculars on a fence post. Notice how much easier it is to split the pair? This is the reason why binocular mounts appeal to so many observers — and the advantage only increases with magnification.

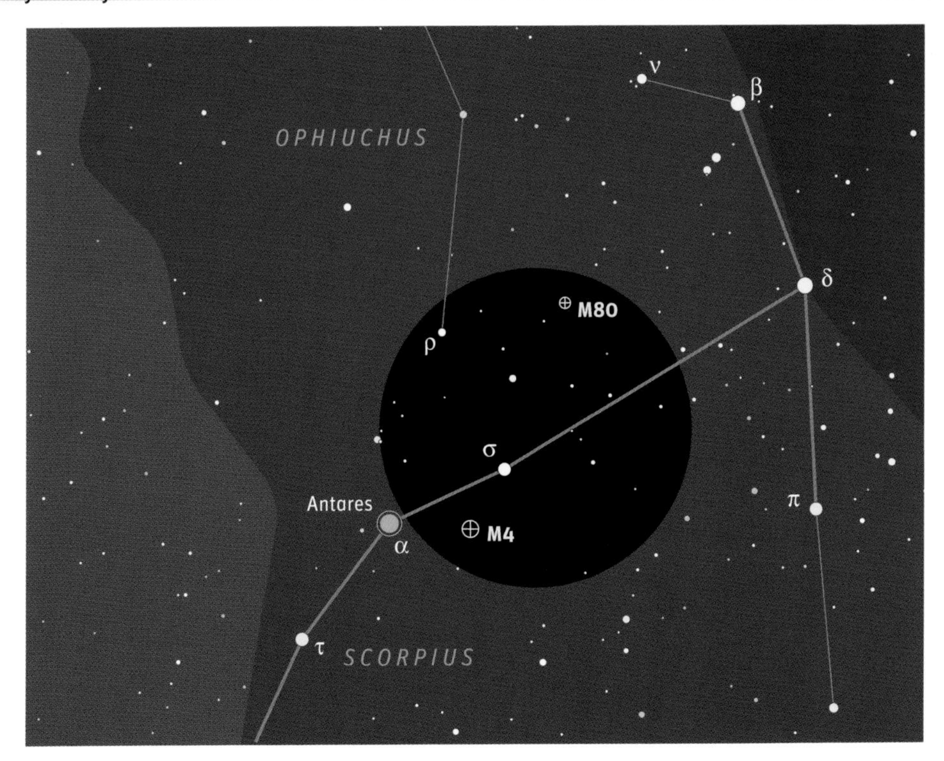

Globular Cluster Season

The Milky Way is home to more than 150 globular star clusters, of which 125 culminate more than 10° above the horizon at midnorthern latitudes. Of course, most of these will be too faint to see in standard binoculars. But how many can an experienced observer using steadily mounted 10 × 50 binoculars under a dark sky actually see? I suspect the answer lies somewhere between 50 and 67, which corresponds to cluster brightness limits of 9th and 10th magnitude, respectively. (Interestingly, all but 2 of the 67 are visible on July evenings.) Observers under excellent skies and using 10× binoculars will bag more than those using 7× binos under less ideal conditions, but these numbers are really only a crude estimate because so many variables come into play.

Of course, a cluster's limiting magnitude is only part of the story — the vast majority of these globulars will also appear stellar in binoculars. The pair of M4 and M80 near Antares illustrates what you're up against. M4 looks obviously nonstellar even in 7× binoculars. Not only does this cluster appear bigger than most other globulars, it lacks a distinctive, compact, starlike core. M4's unusual appearance is both a blessing and a curse. Its large apparent size and diffuse nature make it an obvious standout under dark skies, but in light-polluted conditions the lack of a condensed core can make the cluster difficult to see.

M4's neighbor, M80, is much smaller and fainter and, as such, more typical of the breed. Look closely at this cluster and keep in mind that most other globulars are both smaller and fainter still. Spotting most of them will be challenging and will require patience and careful star-hopping.

So, how many globulars can you find?

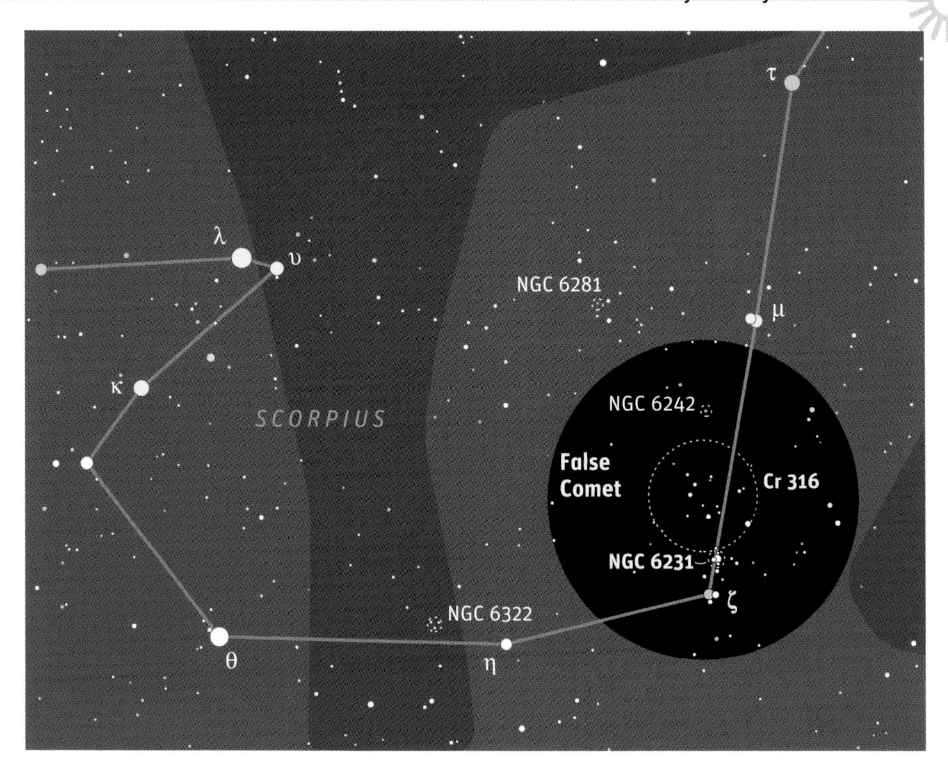

A False Comet

One of the most arresting sights in the summer evening sky is a region of southern Scorpius known as the False Comet. To the unaided eye it really does look like a little comet, with its starlike nucleus, glowing head, and faint, diaphanous tail stretching north. However, binoculars reveal this "comet" to be really a spectacularly rich splash of starlight.

The head of the comet is formed by a neat triangle of bright stars including the wide double star Zeta (ζ) Scorpii, magnitudes 3.6 and 4.7. A half degree north of Zeta is the lovely open cluster NGC 6231 — a tight knot of eight 7th- and 8th-magnitude stars set against a background glow of many fainter ones. From there a string of faint stars leads a degree or two farther north to a pair of large, loose, overlapping groupings known as Collinder 316 and Trumpler 24. To the unaided eye they give the comet its fan-shaped tail, while in binoculars they look like a localized brightening of the Milky Way. Although each piece of the comet can be enjoyed individually, the whole region can easily be taken in with one binocular-size gulp, which is probably the best way to view it — just like a real comet.

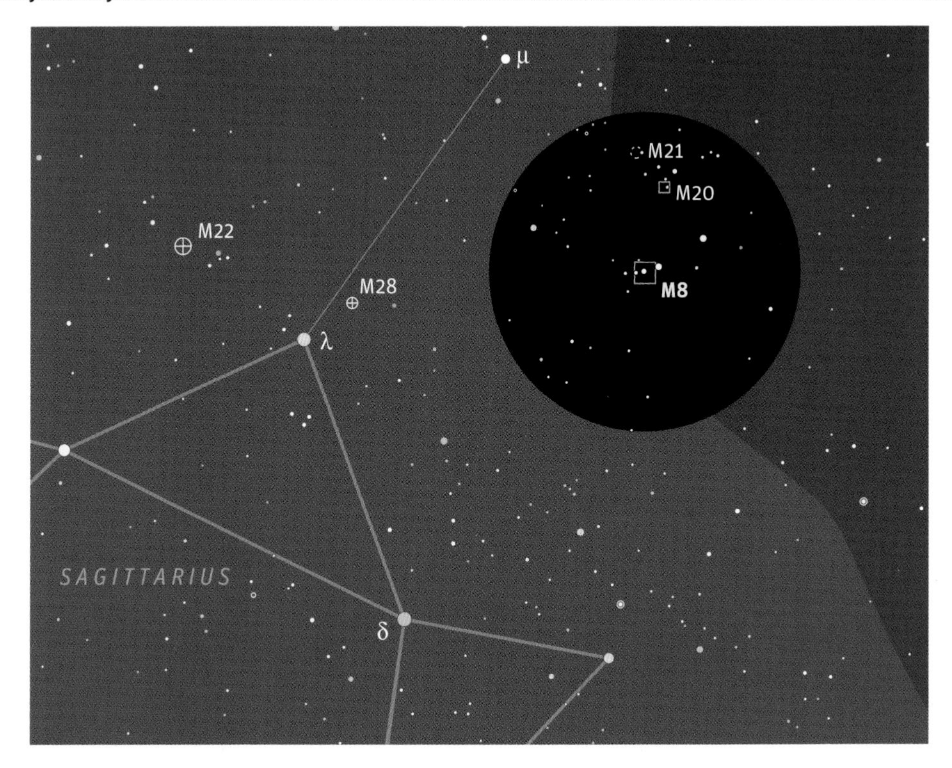

The Lagoon Nebula

This is the best time of year to be a binocular astronomer. The bright core of the Milky Way, with all its attendant treasures, is crossing the meridian, and for many readers the nights are as mild and pleasant as they come. The combination provides the perfect inducement to a night of relaxed viewing from the comfort of a chaise longue. And one of the must-see objects of this time of year is the Lagoon Nebula, M8.

Although one can methodically star-hop to M8 from the spout of the Sagittarius Teapot, the richness of this part of the sky lends itself to a more casual, sweeping approach. You'll know you've reached your destination when you come across a small but conspicuous misty patch of light flanked by a chain of stars aligned east to west.

In his book *The Messier Objects,* Stephen James O'Meara evocatively describes the Lagoon as a "large curdle of galactic vapor." How much of this emission-nebula "vapor" you see will depend, as always, on sky conditions. That said, one April I viewed M8 from my light-polluted suburban backyard while twilight began to brighten the sky, and I had no trouble seeing the Lagoon in my 10 × 30 binoculars. I'm sure the view from your chaise longue will be better.

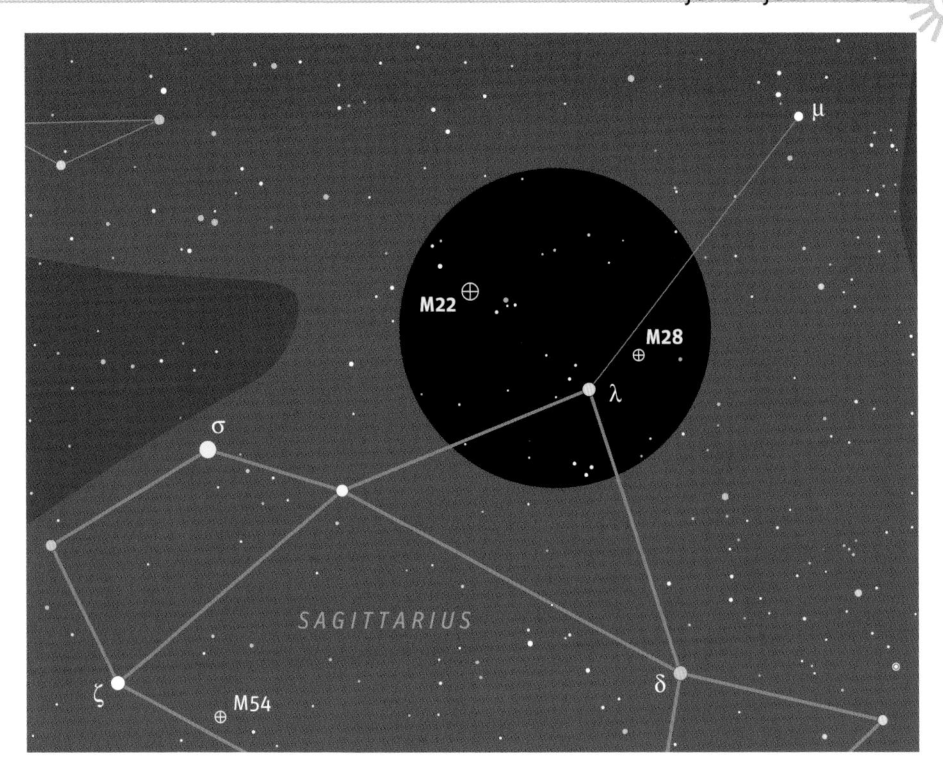

M22: A Gem of a Globular

Most Northern Hemisphere observers harbor the suspicion that all the really good stuff lies just over the southern horizon and can be seen well only from Australia. Although it's true that those of us living at midnorthern latitudes are denied views of the very best globular star clusters in the sky, there are a few gems among the ones that clear our southern horizon — including M22 in Sagittarius.

For binocular viewing, M22 may be the best globular in the entire sky for midnorthern observers. It's bright (magnitude 5.1), large (24′ in diameter), and easy to find just northeast of Lambda (λ) Sagittarii, the star that marks the top of the Teapot. There you will find a cluster that actually looks like more than just a slightly fuzzy star. In 10× binoculars, M22's "globularness" is distinct even in skies compromised by light pollution.

For an even more enjoyable view, shift M22 to left of center so you also get Lambda and the neighboring globular M28 into your view. It's a striking field.

While M13 (described on page 55) is better known, it's actually slightly fainter and smaller than M22. The only other contender I can think of for the title of Best Northern Hemisphere Binocular Globular is M4 in Scorpius, because it too is large, bright, and visually interesting (see page 74). Since all three are visible on a summer evening, why not have a look yourself and see which one you think is the best of all?

AUTUMN

Akira Fujii

SEPTEMBER • OCTOBER • NOVEMBER

PLANETARY NEBULA
GLOBULAR CLUSTER
DIFFUSE NEBULA
OPEN CLUSTER
VARIABLE STAR
GALAXY

ABOUT THE CHARTS:

Each of the star maps in this chapter has been rendered at one of three different scales: the wide-field charts to magnitude 7.5, the medium-scale charts to magnitude 8.0, and the close-up charts to magnitude 8.5. Regardless, the darkened circular area always represents the field of view for typical 10 x 50 binoculars.

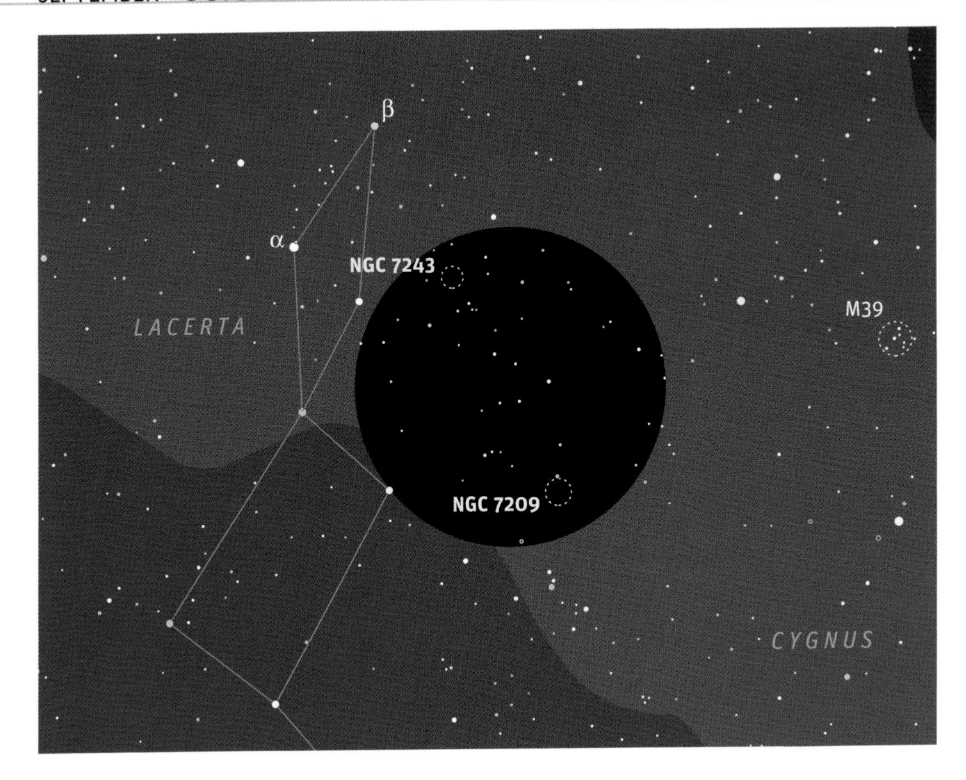

Two Lacerta Clusters

Little Lacerta tends to get lost among its splashier neighbors. Indeed, the constellation is a challenge to trace from suburban locations — its leading light, Alpha (α) Lacertae, is only a little brighter than 4th magnitude. In spite of this, Lacerta does lie along the Milky Way. It occupies the expanse between Cygnus and Cassiopeia and, as such, is home to some pretty rich star fields. In binoculars, the constellation's main stars resemble a miniature replica of neighboring Cassiopeia. And near its Cassiopeia-like W are two open clusters suitable for hunting down under dark skies.

While neither cluster is a showpiece, they are typical of many binocular clusters that simply appear as little areas of localized brightening within the Milky Way. Of the two, NGC 7243 is the more striking object and shows about eight individual stars that form a rough rectangle aligned east to west. With averted vision, several more cluster members sporadically pop into view. Nearby NGC 7209 is a round, diffuse glow with a few faint individual stars occasionally emerging from the background.

What's interesting is that the binocular view is about as good as it gets. I recall one night when I viewed them with my 10 × 30 binoculars. They looked promising enough that I expected them to blossom into striking objects in my 8-inch telescope. To my surprise, the clusters practically vanished. The increased light-gathering power of the 8-inch pulled in so many additional faint stars that the clusters simply blended in with them.

I have seen this effect with other open clusters. In such situations, the limited light-gathering power of ordinary binoculars actually works in your favor.

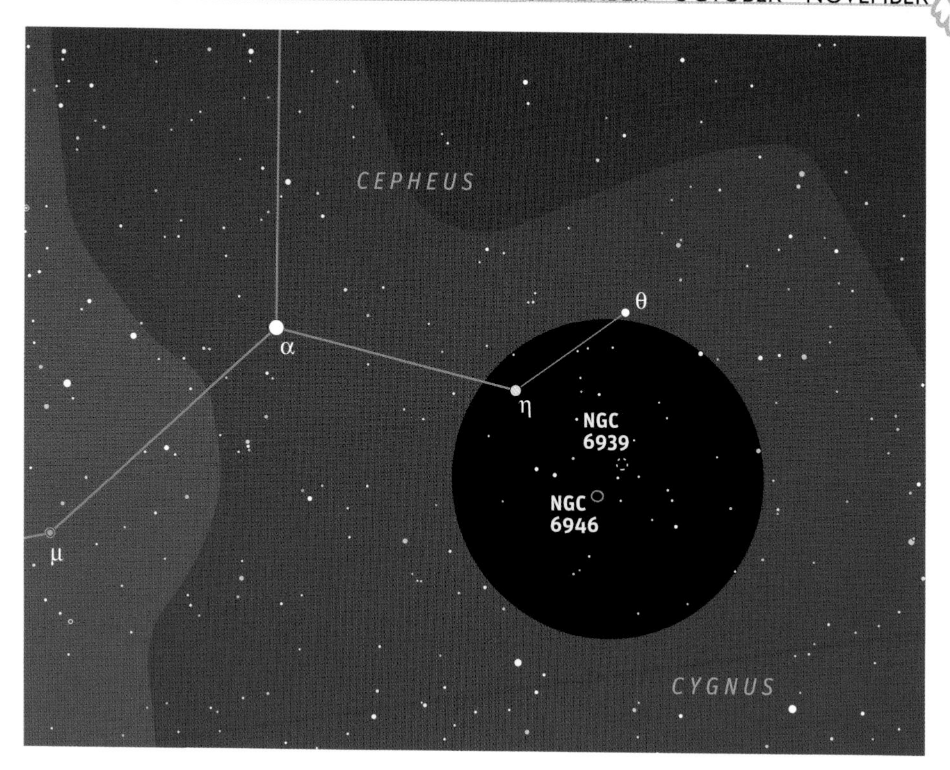

NGC 6939 and 6946

In the blink of an eye we can span light-years — or so it seems. In southwestern Cepheus we find a pair of challenging deep-sky objects separated by only ½° (the apparent diameter of the Moon), yet separated in space by millions of light-years.

The easier of the two to see is NGC 6939, an open cluster 4,000 light-years distant inside our own Milky Way galaxy. Although its apparent neighbor looks rather similar in binoculars, NGC 6946 is in fact a spiral galaxy about 20 million light-years away. As you glance back and forth between these two faint smudges of light, you are shifting your gaze from near to far by a factor of 5,000!

To find this celestial odd couple you will need a dark moonless sky far from city lights. Find Cepheus high in the north. Begin at Alpha (α) Cephei and hop 4° west (nearly the width of a binocular's field of view) to Eta (η) Cephei. Two degrees southwest of Eta you will find the open cluster and galaxy glowing dimly side by side.

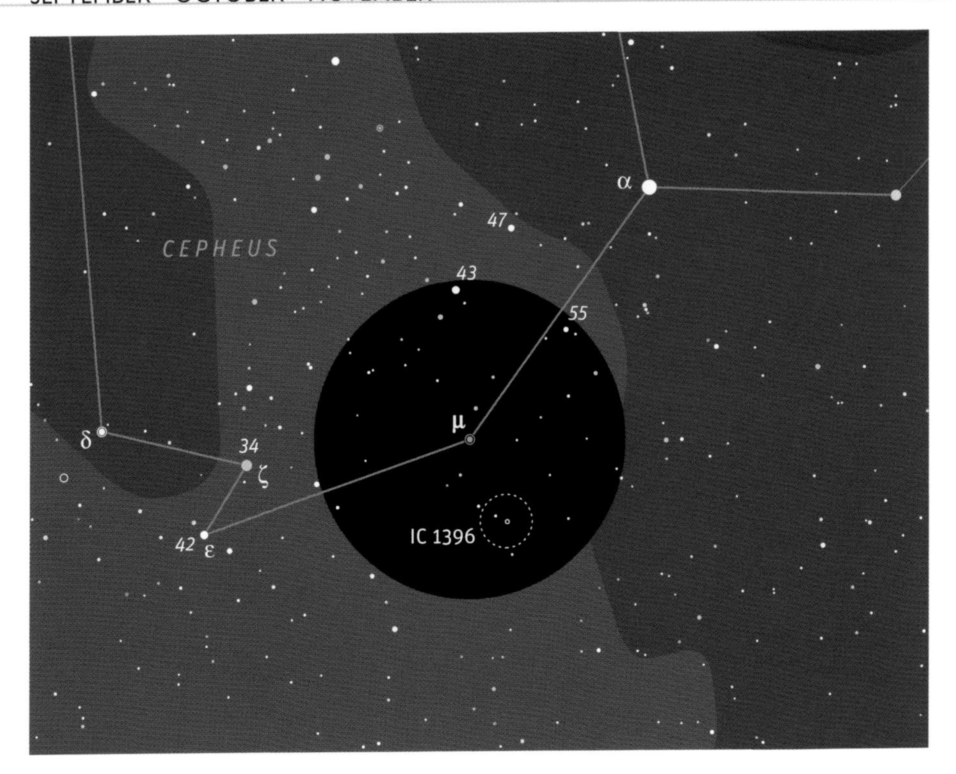

Magnificent Mu Cephei

Individual stars rarely appear on anyone's list of binocular highlights. Let's be honest — even a die-hard observer has trouble sustaining interest in viewing plain ol' stars for their own sake. But there are exceptions — perhaps none more notable than Mu (μ) Cephei.

In many ways, Mu has it all. It's visually interesting and challenges our understanding of star behavior. Mu is a red supergiant, like Betelgeuse in Orion, but even larger. If Mu Cephei lay at the center of our solar system, its outer atmosphere would extend past the orbit of Jupiter! In fact, until recently Mu Cephei was the largest star known.

Mu is known as William Herschel's Garnet Star for its striking red color — indeed, it is the reddest naked-eye star north of the celestial equator. But star colors are generally more subdued than most people believe, so don't expect to see a ruby red beacon shining out from the blackness. That said, even in 10 × 30 binoculars Mu appears distinctly yellowish orange and is easy to identify in a pretty field because of that.

The other aspect that makes Mu visually interesting is that it is a semiregular variable star that can swing from magnitude 3.4 to 5.1 — nearly a fivefold brightness change. (Comparison-star magnitudes in the chart above have their decimal points omitted.) And as if that weren't reason enough to look in on it, consider that its brightness changes are complex and somewhat unpredictable. Data going back to the 1840s suggest two periods of variability: an 850-day cycle and a longer, 4,400-day secondary period. The best way to find out what it's up to right now is to go out and have a look!

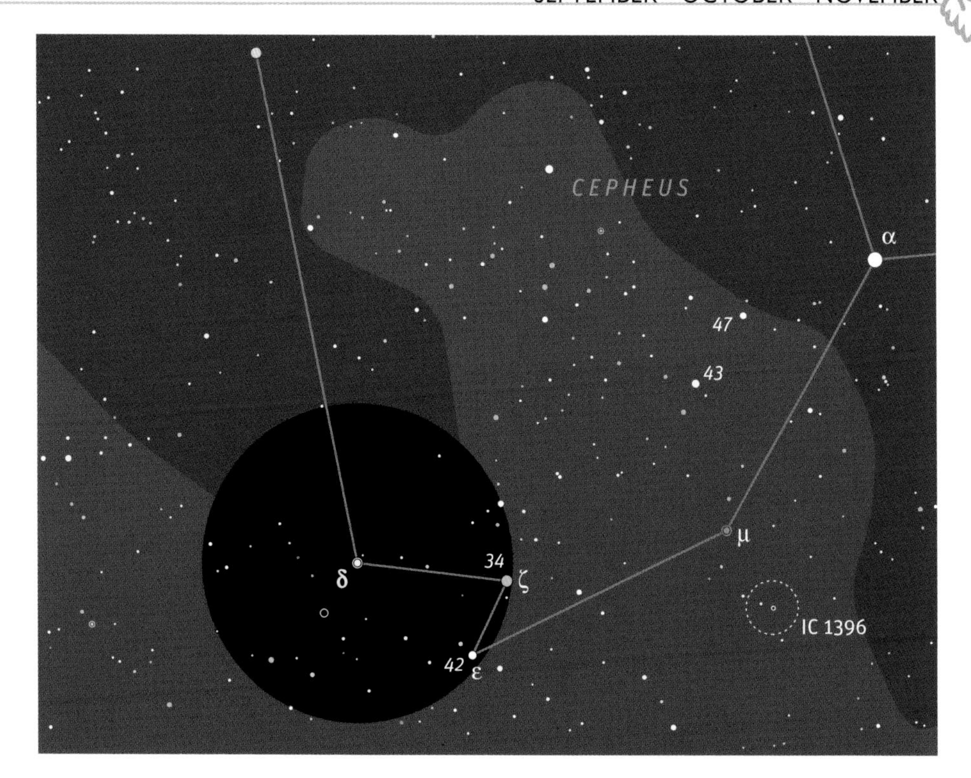

Watching Delta Cephei

Few stars offer as much observing pleasure as Delta (δ) Cephei. Not only is it a well-known variable star, but it is also a striking double. As a double star for binoculars, Delta is a pretty, if somewhat challenging, sight. While I had no trouble splitting the 41″ pair with 15× stabilized binoculars, observers with 7× glasses will likely find the 6.3-magnitude companion quite difficult.

The bright component is most famous as the prototypical Cepheid variable — a class of star that serves as an important "standard candle" indicator of cosmic distances. This particular example takes a little more than 5 days (5.366341, to be precise) to pulse from magnitude 4.4 to 3.5 and back. Binoculars, though not necessary to follow its brightness fluctuations, help city observers when the star drops to the faint end of its range. (In the chart above, magnitudes are given for comparison stars. Decimal points have been omitted, so the star labeled 42, for example, is magnitude 4.2.)

Following Delta's ups and downs is addictive — perhaps after a few nights you will find yourself wanting to delve deeper into the fascinating world of variable-star observing. So if these inconstant stars pique your curiosity, you might consider joining the American Association of Variable Star observers (AAVSO). You can learn more about this organization and about variable stars by visiting their Web site: www.aavso.org.

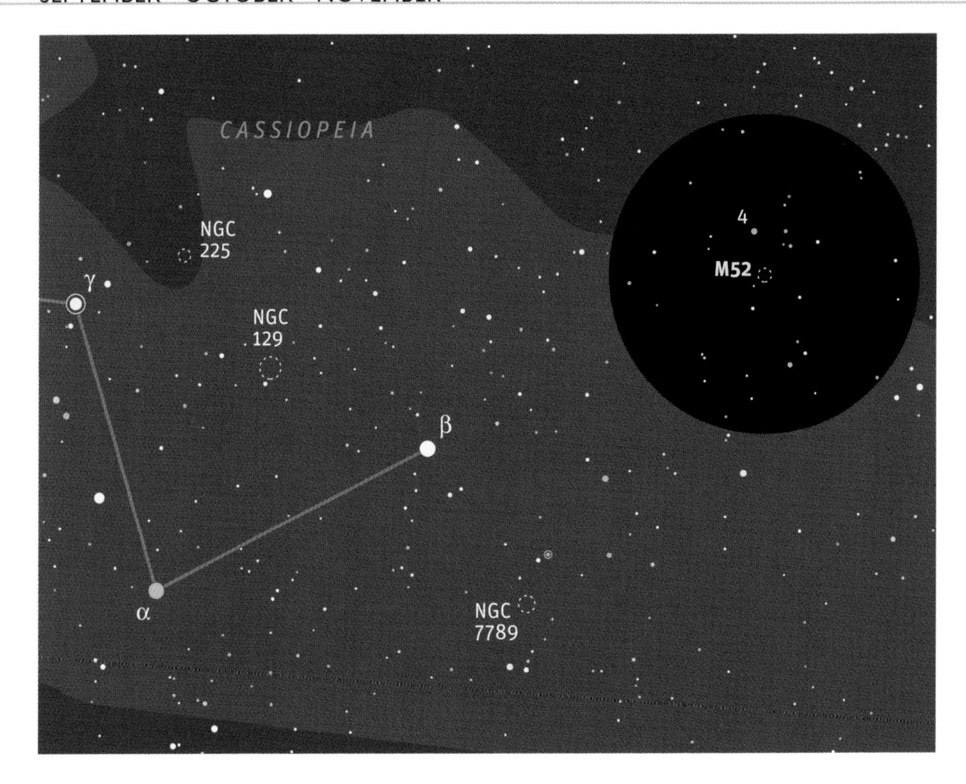

M52 in Cassiopeia

The expanse of Milky Way that stretches from Cygnus to Auriga is rich with star clusters. Eight of the Messier list's open clusters are found here, including M52 in Cassiopeia, riding high in the evening sky at this time of year.

Although it is listed as a reasonably bright object of magnitude 6.9, spotting M52 can be a bit of a challenge, depending on the quality of your skies. It's certainly more difficult than picking out a 7th-magnitude star, since its light is spread out into a diffuse smudge covering about 12′.

Fortunately, M52's location is easy to find. Simply extend a line from Alpha (α) through Beta (β) Cassiopeiae onward a distance equal to the separation of the two stars, and there you are. You should see M52's glow only 0.8° south of 5th-magnitude 4 Cassiopeiae. From my suburban backyard I can distinctly pick out only one individual from the haze of fainter cluster stars in 10 × 30 binoculars. How about you?

Also, while you are in the neighborhood, check out NGC 7789 (see opposite) and the attractive curving row of 6th- and 7th-magnitude stars less than 1° west of 4 Cassiopeiae.

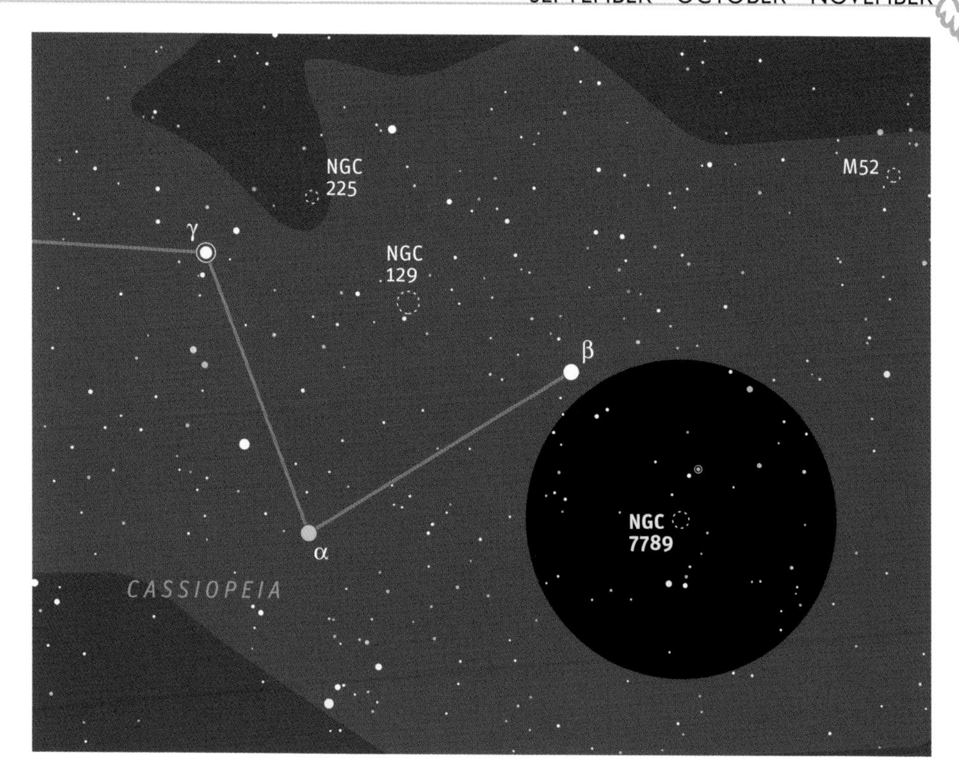

Galactic Cluster NGC 7789

Many amateur astronomers assume that Charles Messier's 18th-century list of comets-that-weren't represents the sky's best and brightest deep-sky objects. But for one reason or another Messier left a lot of wonderful objects off his list, including some open clusters that make for fine binocular observing. One particularly lovely non-Messier cluster is NGC 7789. It can be found off the western end of the W of Cassiopeia, about a half binocular field southwest of Beta (β) Cassiopeiae.

This cluster is one of the unsung gems of the autumn Milky Way, which runs from northern Cygnus through Cassiopeia, Perseus, and into Auriga, where it thins out. I often "rediscover" NGC 7789 in my casual sweeps of this region of sky. Although star-hopping to the cluster will get you there fastest, I find the scenic route more enjoyable.

NGC 7789 appears as a reasonably conspicuous round, hazy glow set against a rich star field peppered with 7th- and 8th-magnitude stars. Don't expect ordinary binoculars to reveal individual stars in the cluster itself. Although NGC 7789 contains roughly 1,000 stars, they are uniformly faint.

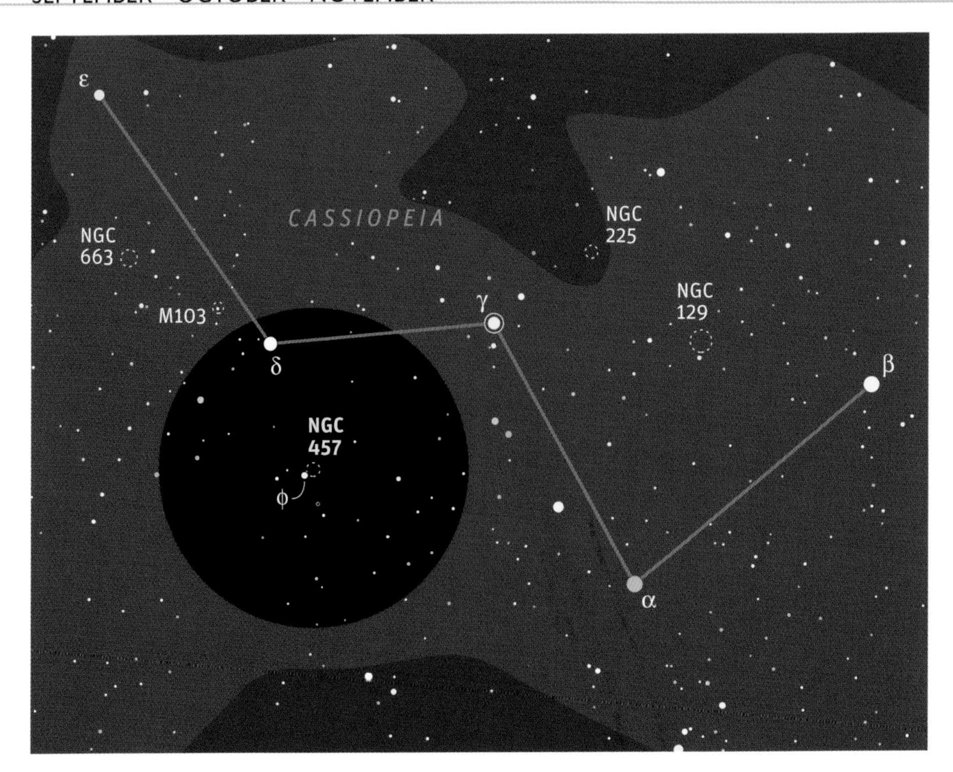

The E.T. Cluster

The whole of Cassiopeia is a binocular-observer's paradise. Few stretches of the northern Milky Way boast so many interesting clusters and star-filled vistas. Of course, it helps that the constellation itself is so easy to navigate — even novice stargazers should be able to find their away around Cassiopeia's distinctive W shape.

Located 2° south-southwest of Delta (δ) Cassiopeiae is one of the region's most attractive sights: NGC 457, otherwise known as the E.T. Cluster because of its resemblance to the charming alien featured in the 1982 movie, *E.T. — The Extraterrestrial.* At this time of year, E.T. is standing on its head. Phi (φ) Cassiopeiae and its 7th-magnitude companion serve as the alien's eyes, glowing in the darkness. Strings of stars suggest the body and legs, while a pair of curving rows mark E.T.'s arms, one of which is raised as if pointing to the heavens. Patience, dark skies, and tripod-mounted binoculars that magnify at least 10× are needed to pick out these features of E.T.'s anatomy. Lower powers will still show the cluster as a hazy, elongated glow just to the side of Phi and its companion.

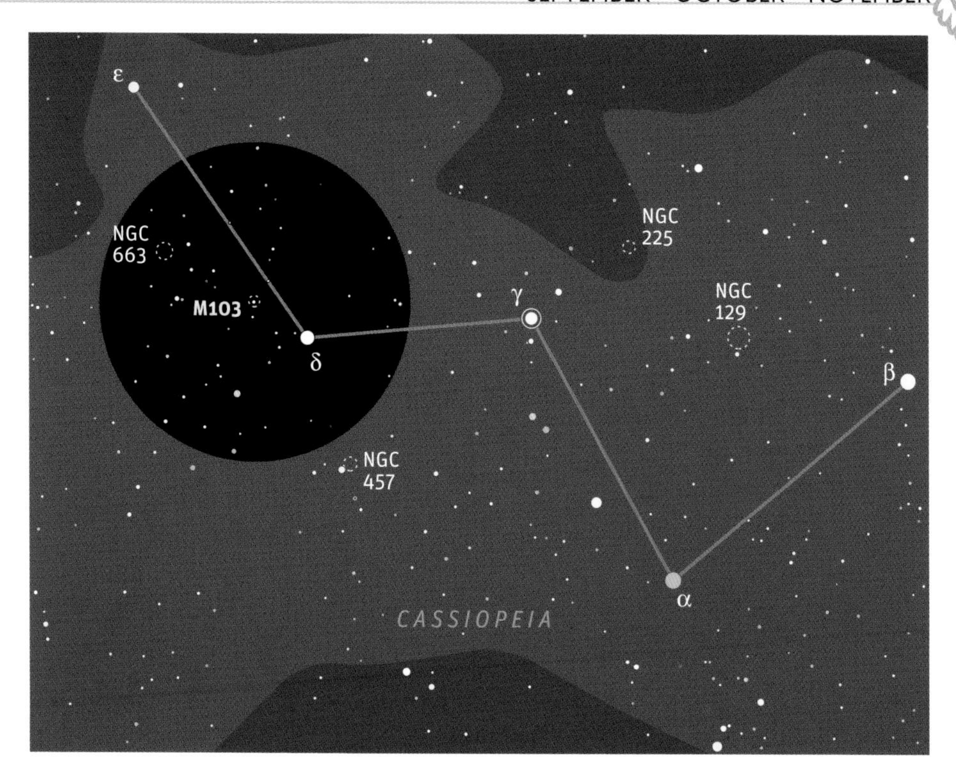

Open Cluster M103

The Cassiopeia section of the Milky Way is chock-a-block with open clusters, which tend to be the most interesting targets for low-magnification, wide-field instruments in general. Indeed, the classic three-volume *Burnham's Celestial Handbook* lists no fewer than 26 open clusters in Cassiopeia, more than for any other constellation. Yet Messier noted only two open clusters here: M52 (page 84) and M103.

If you center your binoculars on Delta (δ) Cassiopeiae you are in downtown Clusterville — M103 shares the field of view with NGC 663 as well as with several other, fainter clusters. Being the brightest, and a scant 1° northeast of Delta, M103 is easy to locate. Binoculars reveal it as a tight knot of stars with three conspicuous members that form one side of a tiny equilateral triangle. Magnification definitely helps show this aspect, so use binoculars that magnify 10× or greater if you have them. Even in light-polluted suburban skies, M103 has enough bright stars to attract the eye's attention. Although it may contain as many as 100 individual suns, few of these cluster members are bright enough to be visible in binoculars under even the darkest skies.

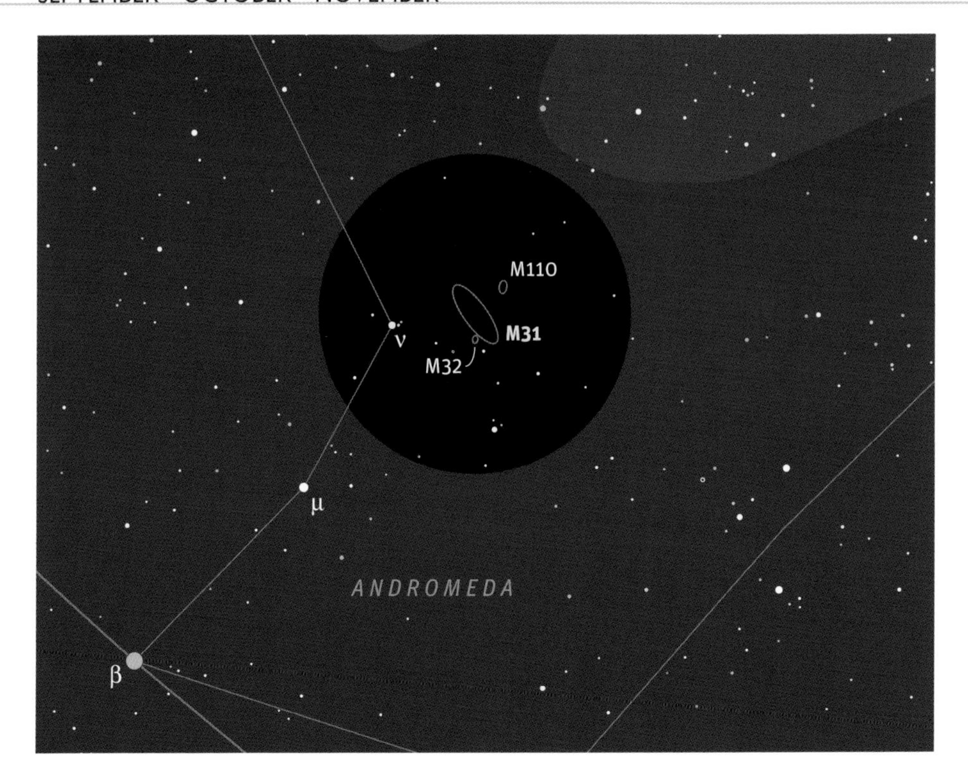

Magnificent M31

There is no disputing that the most impressive galaxy for binocular observers is M31, the Great Andromeda Galaxy. Excluding the Milky Way and the Large and Small Magellanic Clouds, M31 is the brightest and biggest galaxy either north or south of the celestial equator. It's also one of a handful of objects that is as rewarding to view through binoculars as with a telescope.

Located at the end of a string of stars consisting of Beta (β), Mu (μ), and Nu (ν) Andromedae, the galaxy is easy to see without optical aid as long as your sky is reasonably dark. Its impressiveness is because of its proximity to us. At a distance of about 2.5 million light-years, M31 is the nearest large galaxy to our own. It's also big — a little larger than the Milky Way.

How the galaxy appears in your binoculars will depend on the quality of your skies. From the city, M31 looks very much like a small, tailless comet — only the brightest part of its nucleus shines through. However, from a dark-sky location, this bright core blooms into a nearly symmetrical, dim ellipse that spans half the binocular field. Take your time to enjoy the view, and use *averted vision* (look slightly to one side of the object) to see how far you can trace the galaxy's extent. M31's glowing expanse, set against a peppering of Milky Way foreground stars, is one of the binocular sky's most compelling vistas.

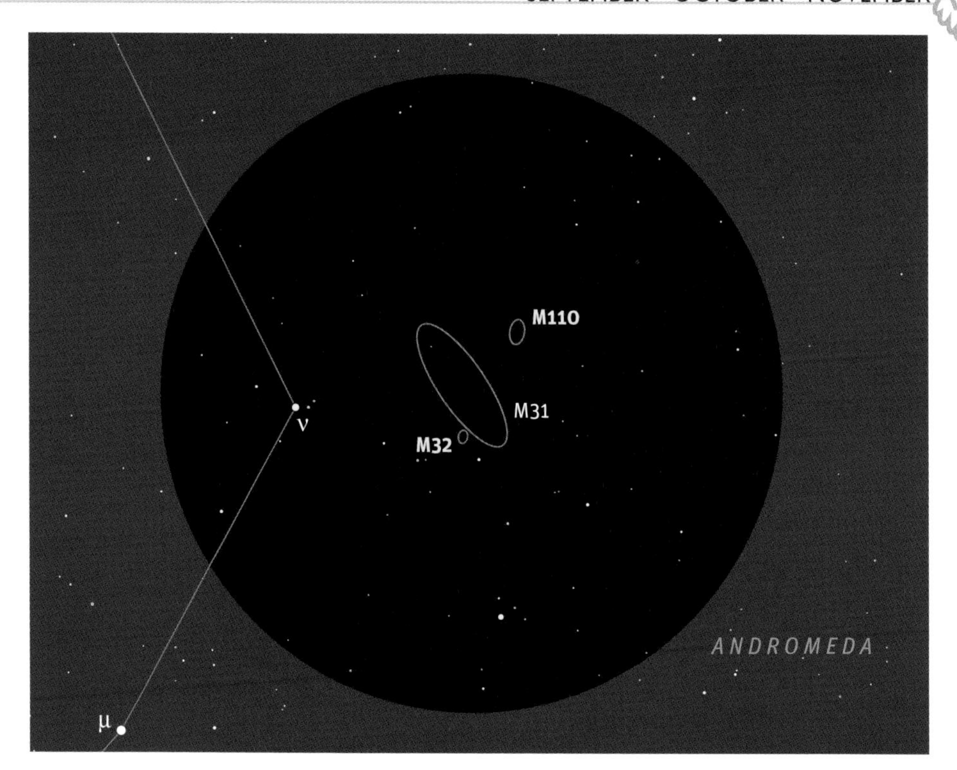

Andromeda's Companions

Any list of the night sky's showpiece objects will surely include the Andromeda Galaxy, M31. Perhaps because it is such a well-known deep-sky jewel, its two companion galaxies receive little attention.

Due south of the main galaxy's bright nucleus lies the dwarf elliptical M32. This object causes difficulty for binocular observers not so much because it's faint (magnitude 8.1) but, rather, because at the low magnification provided by typical binoculars it can be tricky to distinguish from a star. Fortunately, it's located only 0.2° northeast of a 7th-magnitude star that serves as a handy comparison object. Although in 10× binoculars M32 appears tiny, next to its stellar neighbor the small galaxy's fuzzy appearance is obvious — it looks like a star that's slightly out of focus.

The Andromeda Galaxy's other companion is M110, an elliptical galaxy northnorthwest of M31. While M32 shows up in moderately light-polluted conditions, dark skies are required to glimpse M110. Compared with M32, this galaxy is only slightly fainter (magnitude 8.9), but its diffuse character makes it more difficult to see. In my 10 × 50 binoculars I find that averted vision helps to make its indistinct glow easier to detect.

While neither M32 nor M110 even comes close to matching the visual impact of the Andromeda Galaxy, each is an interesting object in its own right. So, the next time you stop in on M31, take a few minutes to appreciate its neglected companions.

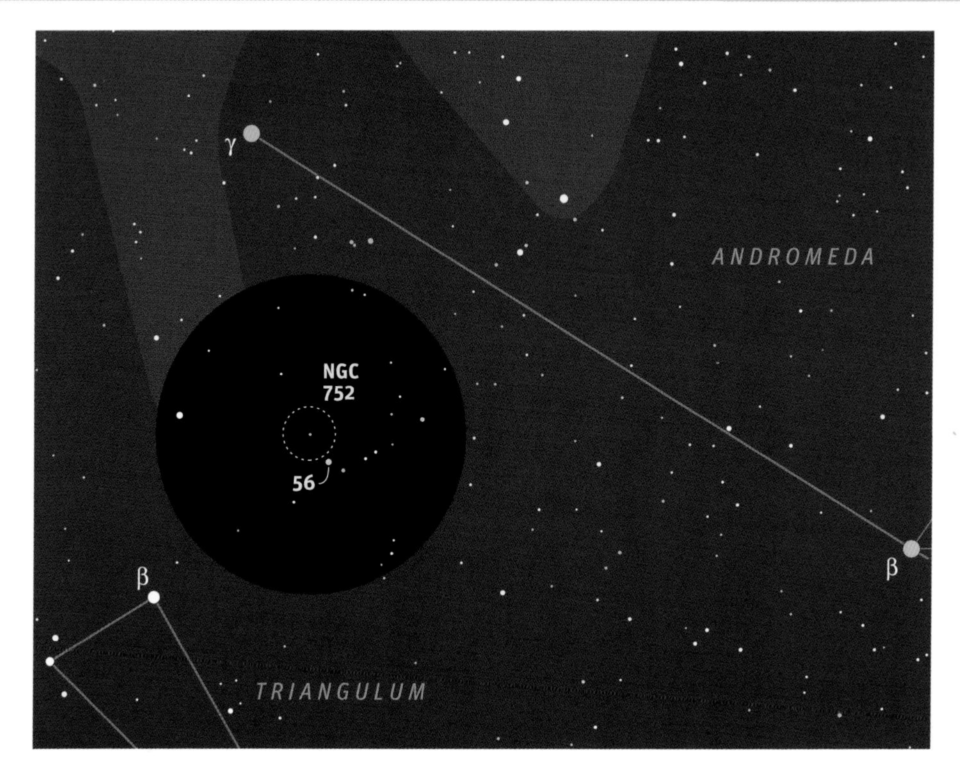

Ancient Cluster NGC 752

Never heard of NGC 752? That's hardly surprising — it's easy to overlook a cluster like this when it shares the evening sky with the magnificent galaxy M31 and the spectacular Double Cluster in Perseus. Still, NGC 752 does have its charm and is well worth seeking out.

The most remarkable fact about NGC 752 is that it is very old. Most open clusters have ages measured in tens or hundreds of millions of years. However, astronomers peg NGC 752 at about 2 billion years old. Its relatively ancient nature is subtly apparent in binoculars. Cluster stars tend to spread out as they age, and the sparse appearance of NGC 752 is at least in part a result of this. Indeed, it's something of an observing challenge to judge the location of the cluster's center.

One of the best views that I've had of NGC 752 came one summer while I was camping in the Canadian Rockies. The cluster was faintly visible to the unaided eye, and 10 × 30 image-stabilized binoculars showed it very well. With averted vision I could see about a dozen faint cluster members popping in and out of view, but it was difficult to tell where the cluster ended and the field stars began. The combination of cluster and field stars gave the impression of an object close to 2½° across, though the cluster's cataloged size is less than half that.

As a bonus attraction, have a look at the binocular double star 56 Andromedae, lying near the cluster's southwest edge. The pair consists of 6th-magnitude stars separated by a generous 200″ — an easy target for observers under less-than-pristine skies.

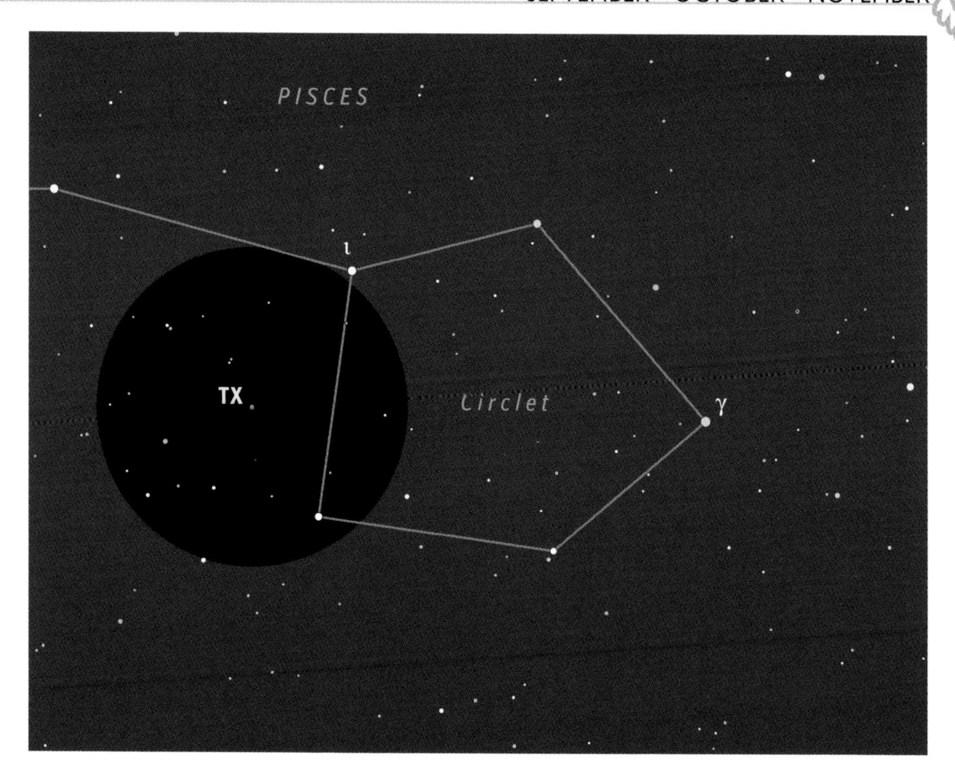

TX Piscium

Just south of the Great Square of Pegasus lies a group of stars in Pisces known as the Circlet. Our chart above plots the Circlet's five brightest stars, but under pristine skies you should be able to see two more — giving it a more oval shape. The easternmost star of these two additions is the carbon star TX Piscium, which varies slightly from magnitude 4.8 to 5.2.

TX (also listed as 19 Piscium) is among the reddest stars known thanks to the carbon in its atmosphere, which acts as a red filter blocking shorter (blue) wavelengths of light. In spite of this, the star's color is rather subdued. Through binoculars TX has a deep golden orange hue that is apparent enough when you are looking for it but might be easy to miss while you're casually scanning the region.

This carbon star demonstrates one aspect of visual astronomy that is a frequent source of surprise for beginning skygazers — the beauty of the universe is far more subtle than is usually expected. Extremes of color and brightness are rare. And though TX can legitimately be called a "red" star, such descriptions are, to some extent, in the eye of the beholder.

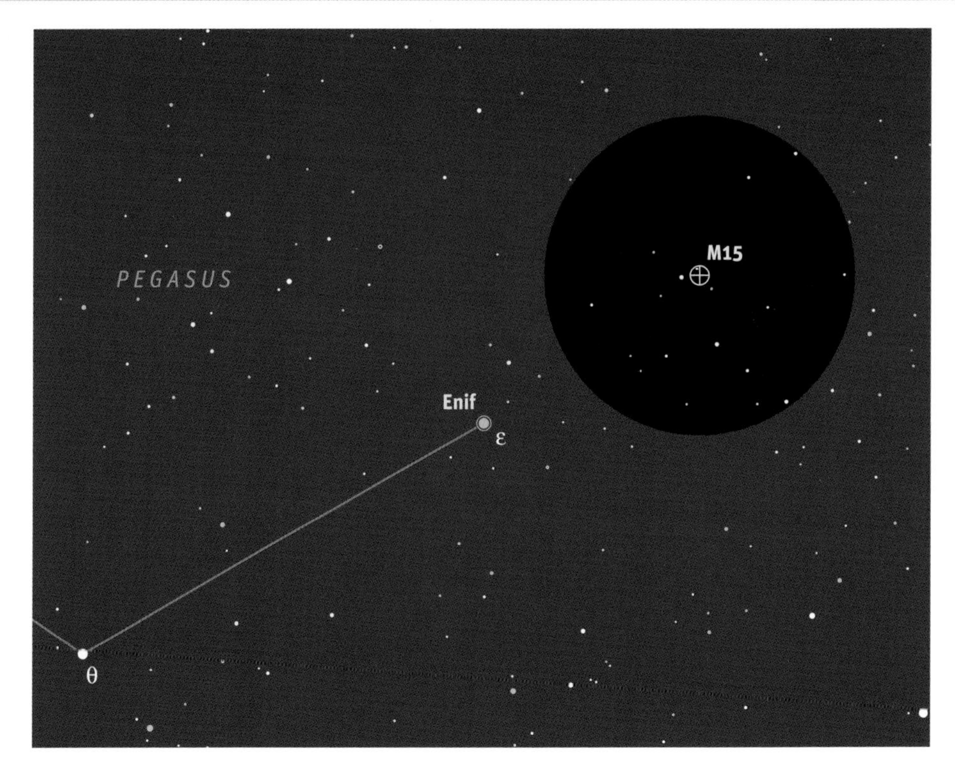

A Farewell to Summer

On page 40 I remarked that the arrival of the globular cluster M3 in the evening sky signifies that summer in the Northern Hemisphere is just around the corner. As M3 marks the leading edge of summer's swarm of bright globular clusters, so does M15 tag the trailing edge that signals summer's end. These seasonal bookends share a number of features, particularly for the binocular observer.

Riding high in the autumn evening sky, M15 has roughly the same brightness as a 6th-magnitude star — just like M3. This means that it too is bright enough to be seen in binoculars under just about any sky regardless of light pollution. Both clusters appear as small, almost starlike glows. However, M15 differs from M3 in one important way — it's a snap to find.

Almost due west of the Great Square of Pegasus is the 2.4-magnitude star Enif, Epsilon (ε) Pegasi. Moving Enif to the southeast edge of your binocular field of view will bring M15 in on the northwest edge. Together the globular and the autumn-gold star make a striking pair.

Compared with galaxies, open clusters, and even planetary nebulae, globular clusters are among the sky's rarest birds. The latest tally for globular clusters belonging to the Milky Way lists only 154 objects, and those as spectacular as M15 are rarer still — only 11 globulars shine brighter.

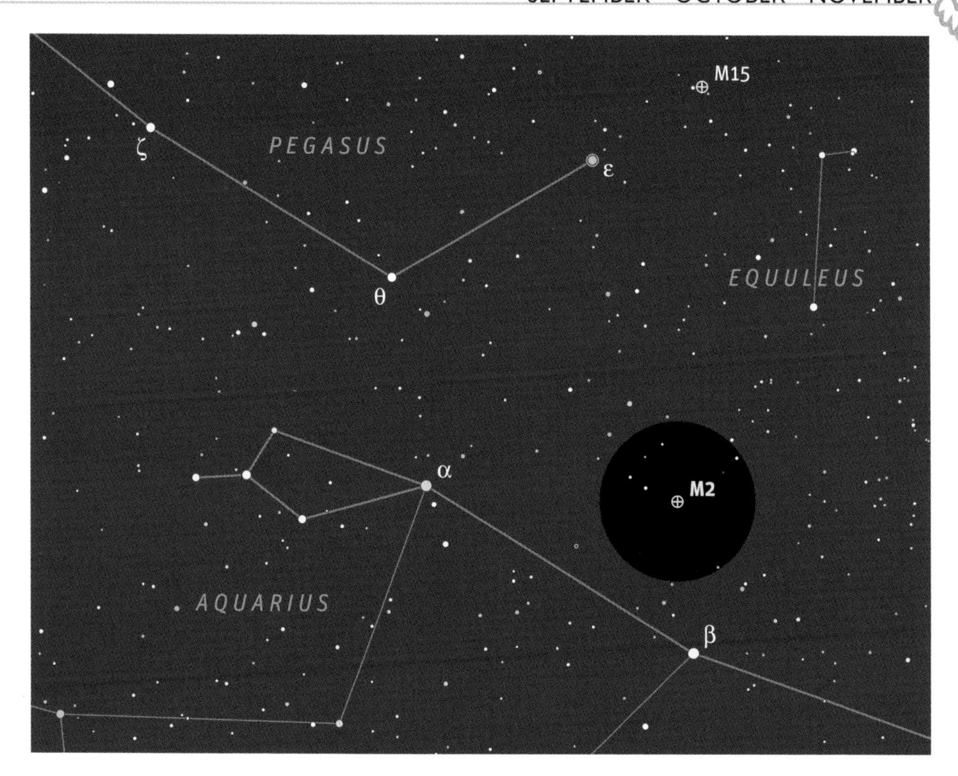

Messier 2 Shining Through

City-bound and suburban observers have it tough when it comes to exploring the wonders of the deep sky. Light pollution robs the darkness from night, making many faint and fuzzy delights nearly impossible to view. However, a bright sky does not affect all deep-sky objects equally — some types survive relatively unharmed.

Like many of the best globular star clusters, M2 is both small and bright, giving it a high surface brightness. Because its luminosity is concentrated in a small area, in binoculars it looks a lot like a 6.4-magnitude star that is ever so slightly out of focus. While this starlike appearance makes it easy to *see,* it makes it more difficult to *identify*. Fortunately, M2 lies just far enough from the crowded star fields of the Milky Way that there are few similarly bright points of light to confuse with it. If you place Beta (β) Aquarii at the bottom of the binocular field of view, you should be able to pick out the cluster even in a bright urban sky. And while you're in the neighborhood, slide north to have a look at M15, described opposite. Both clusters have the same brightness and appear about the same size, but I'd be willing to be you'll find one looks slightly more conspicuous than the other.

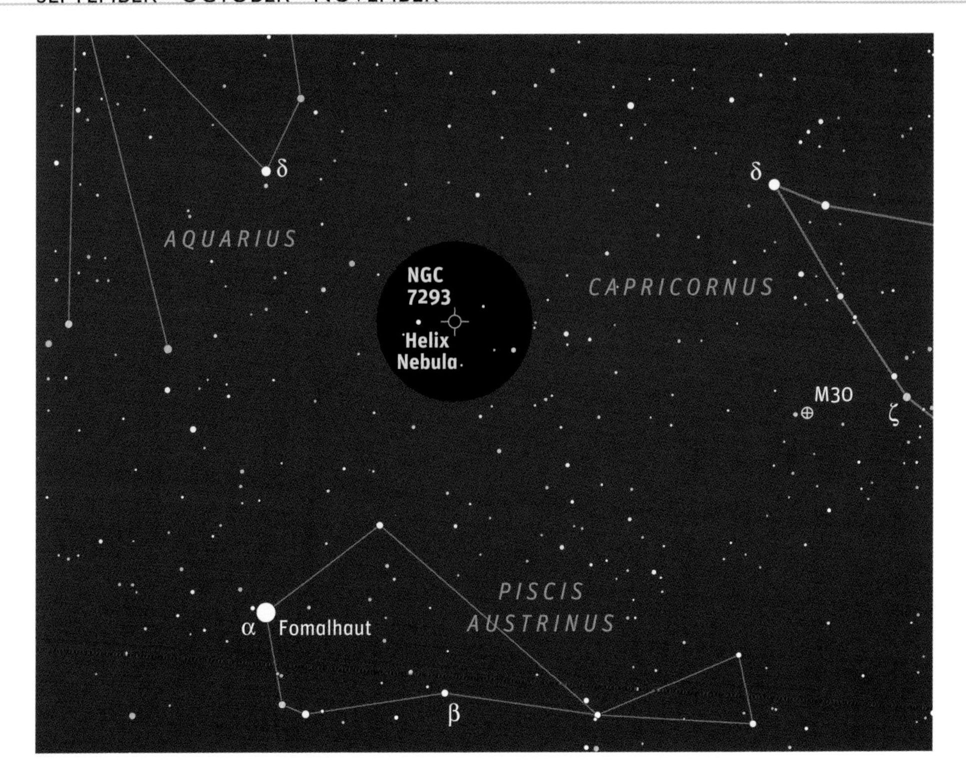

The Ghostly Helix Nebula

Most planetary nebulae are faint, starlike objects even in large amateur telescopes. A handful of famous examples, like the Ring Nebula in Lyra (page 61), appear as small glowing disks. Rarest of all are those planetaries big and bright enough to be interesting binocular targets. NGC 7293, the Helix Nebula, is one such object.

The Helix is located in a particularly barren stretch of sky in the constellation Aquarius. You can best find it by starting off at Delta (δ) Capricorni and heading southeast toward the region's brightest star, Fomalhaut. About midway between is the planetary nebula's field.

With its cataloged size of nearly 13′ (more than 10 times the diameter of the Ring Nebula) and a magnitude of 7.3, you might not expect the Helix to be a particularly tough find. But be warned — its luminosity is spread over a large area, rendering it a difficult, low-surface-brightness object. That said, I was able to spot its round glow without too much difficulty in 10 × 30 image-stabilized binoculars in the very dark skies of Mount Kobau in British Columbia, Canada. Binoculars with more magnification might be able to show the Helix even in less pristine skies.

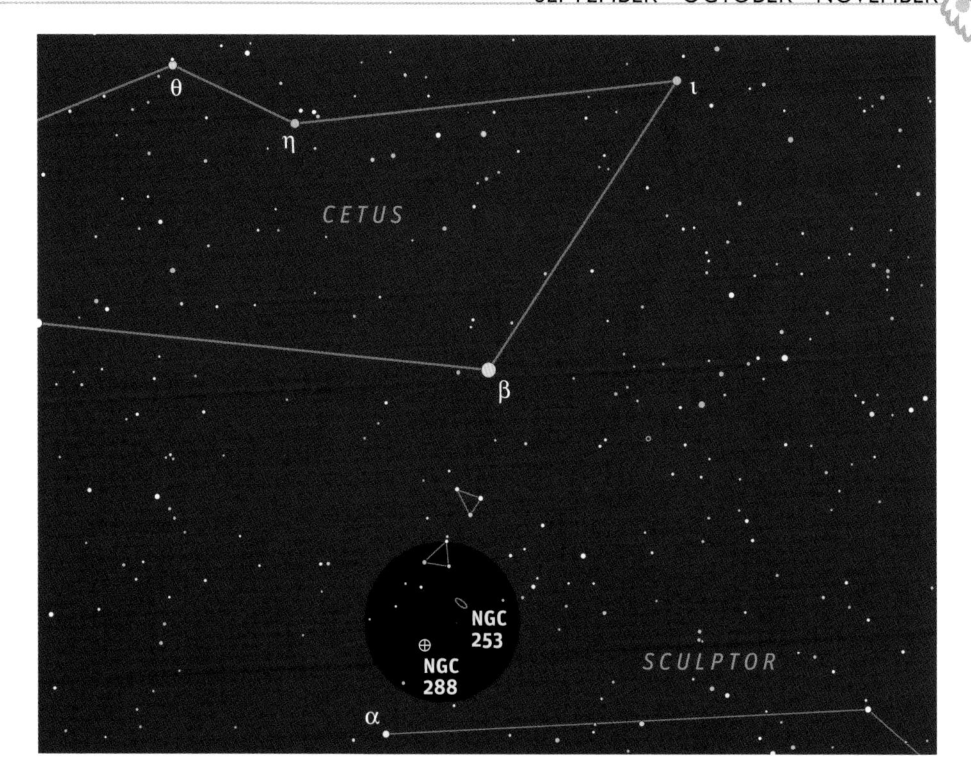

Challenging NGC 253 and 288

Some binocular observers enjoy hunting down deep-sky objects at the limits of visibility. Of course, the threshold for "challenging" in binoculars is different than in telescopes, but the thrill of capturing an elusive target is the same. Autumn evenings provide two objects to test your observing skill and sky conditions: edge-on spiral galaxy NGC 253 and globular cluster NGC 288.

Begin with the easier of the two, the galaxy NGC 253. Two distinctive triangles of 5th- and 6th-magnitude stars will guide you from Beta (β) Ceti to its position. Under a clear, dark, moonless sky even 7 × 50 binoculars will readily show NGC 253's elongated shape.

If you manage to find NGC 253 without too much trouble, see if you can locate much smaller NGC 288, less than 2° to the southeast. Although the cluster is large enough to appear nonstellar in 10× binoculars, it is sufficiently dim that it won't jump out of the field at you. Use averted vision to draw it out. With 10 × 30 binoculars I was able to glimpse NGC 288 from a very dark site, but don't be discouraged if this one gets away — your skies may simply not be good enough.

Const.	Object	Magnitude	Type	Page
And	M31 (Andromeda Galaxy)	4.3	Galaxy	88
And	M32	8.1	Galaxy	89
And	M110	8.9	Galaxy	89
And	NGC 752	5.7	Open Cluster	90
Aql	Barnard's E	n/a	Nebula	65
Aqr	M2	6.4	Globular Cluster	93
Aqr	NGC 7293 (Helix)	7.3	Nebula	94
Aur	M36	6.0	Open Cluster	24
Aur	M37	5.6	Open Cluster	24
Aur	M38	6.4	Open Cluster	24
Boo	Delta Bootis	3.6, 7.9	Double Star	42
Boo	Mu Bootis	4.3, 6.5	Double Star	42
Boo	Nu Bootis	5.0, 5.0	Double Star	42
Cam	Kemble's Cascade	n/a	Asterism	16
Cam	NGC 1502	6.9	Open Cluster	16
Cas	M52	6.9	Open Cluster	84
Cas	M103	7.4	Open Cluster	87
Cas	NGC 457 (E.T. Cluster)	6.4	Open Cluster	86
Cas	NGC 7789	6.7	Open Cluster	85
Cep	Delta Cephei	3.5 – 4.4, 6.3	Variable/Double Star	83
Cep	Mu Cephei	3.4 – 5.1	Variable Star	82
Cep	NGC 6939	7.8	Open Cluster	81
Cep	NGC 6946	9.7	Galaxy	81
CMa	M41	4.5	Open Cluster	28
Cnc	Iota Cancri	4.0, 6.5	Double Star	44
Cnc	M44 (Beehive)	3.1	Open Cluster	45
Cnc	Rho Cancri	5.9, 6.3	Double Star	44
Com	Melotte 111	n/a	Open Cluster	41
CrB	R Corona Borealis	5.8 – 14.8	Variable Star	43
CVn	M3	5.9	Globular Cluster	40
CVn	M51	8.9	Galaxy	37
CVn	M94	8.2	Galaxy	39
CVn	M106	9.1	Galaxy	38
Cyg	61 Cygni	5.2, 6.0	Double Star	57
Cyg	79 Cygni	5.7, 7.0	Double Star	57
Cyg	B168	n/a	Nebula	59
Cyg	M39	4.6	Open Cluster	58
Cyg	Mu Cygni	4.4, 7.0	Double Star	57
Cyg	Omicron^1 Cygni	3.8, 4.8, 7.0	Double Star	56
Dra	Nu Draconis	4.8, 4.9	Double Star	54
Gem	M35	5.1	Open Cluster	25
Gem	NGC 2158	8.6	Open Cluster	25
Her	M13 (Great Hercules Cluster)	5.8	Globular Cluster	55
Hya	M48	5.8	Open Cluster	48
Hya	U Hydrae	5 – 6	Variable Star	49
Hya	V Hydrae	6 – 10	Variable Star	49
Lac	NGC 7209	7.7	Open Cluster	80
Lac	NGC 7243	6.4	Open Cluster	80
Leo	NGC 2903	9.6	Galaxy	46
Leo	Regulus	1.4, 8.1	Double Star	47

Const.	Object	Magnitude	Type	Page
Leo	Tau Leonis	5.0, 7.5	Double Star	47
Lyr	Epsilon Lyrae (Double-Double)	5.0, 5.2	Double Star	60
Lyr	M57 (Ring)	8.8	Nebula	61
Lyr	Vega	0.6	Star	60
Lyr	Zeta Lyrae	4.3, 5.6	Double Star	60
Mon	M50	5.9	Open Cluster	29
Oph	IC 4665	4.2	Open Cluster	69
Oph	M10	6.6	Globular Cluster	70
Oph	M12	6.7	Globular Cluster	70
Oph	NGC 6633	4.6	Open Cluster	68
Oph	Rho Ophiuchi	5.0, 6.8, 7.3	Double Star	71
Ori	Betelgeuse	0.5	Star	26
Ori	M42 (Orion Nebula)	4.0	Nebula	27
Ori	NGC 1981	4.2	Open Cluster	27
Ori	Struve 747	4.8, 5.7	Double Star	27
Peg	M15	6.2	Globular Cluster	92
Per	Algol	2.1 – 3.4	Variable Star	20
Per	Alpha Persei Association	n/a	Open Cluster	18
Per	Double Cluster	5.3, 6.1	Open Cluster	17
Per	M34	5.2	Open Cluster	19
Psc	TX Piscium	4.8 – 5.2	Variable Star	91
Pup	M46	6.1	Open Cluster	30
Pup	M47	4.4	Open Cluster	30
Pup	NGC 2451	3.5	Open Cluster	31
Pup	NGC 2477	5.0	Open Cluster	31
Scl	NGC 253	8.0	Galaxy	95
Scl	NGC 288	8.1	Globular Cluster	95
Sco	18 Scorpii	5.5	Star	72
Sco	False Comet	n/a	Asterism	75
Sco	M4	5.6	Globular Cluster	74
Sco	M80	7.3	Globular Cluster	74
Sco	Nu Scorpii	4.4, 6.5	Double Star	73
Sct	M11	5.8	Open Cluster	66
Ser (Caput)	M5	5.7	Globular Cluster	51
Ser (Cauda)	IC 4756	4.6	Open Cluster	67
Ser (Cauda)	Theta Serpentis	4.5, 5.4	Double Star	67
Sge	M71	8.2	Globular Cluster	62
Sgr	M8 (Lagoon)	5.0	Nebula	76
Sgr	M22	5.1	Globular Cluster	77
Tau	Hyades	n/a	Open Cluster	22
Tau	M45 (Pleiades)	n/a	Open Cluster	21
Tau	NGC 1647	6.4	Open Cluster	23
UMa	M81	7.8	Galaxy	35
UMa	M82	9.2	Galaxy	35
UMa	M101	8.2	Galaxy	36
UMi	Engagement Ring	n/a	Asterism	34
Vir	M104	8.0	Galaxy	50
Vul	Cr 399 (Coathanger)	n/a	Asterism	64
Vul	M27 (Dumbbell)	7.3	Nebula	63

SEPTEMBER • OCTOBER • NOVEMBER

WHEN TO USE THIS STAR MAP:

Early September:	Midnight
Late September:	11 p.m.
Early October:	10 p.m.
Late October:	9 p.m.
Early November:	8 p.m.
Late November:	7 p.m.

This star chart is most accurate if used within an hour or so of the times listed and is plotted for observers located between 30° and 50° north latitude. All times are standard time; if daylight-saving time is in effect, add one hour.

To use this chart, hold it in front of you and rotate it so that the yellow label corresponding to the direction you are facing is positioned at the bottom right-side up. The stars in the sky should match those depicted on the chart. The farther up from the map's edge they appear, the higher they'll be shining in your sky.

The circled numbers on the chart refer to the pages where objects in that region of the sky are described in this book. The numbers highlighted in red indicate the objects best seen at the times and dates listed above.

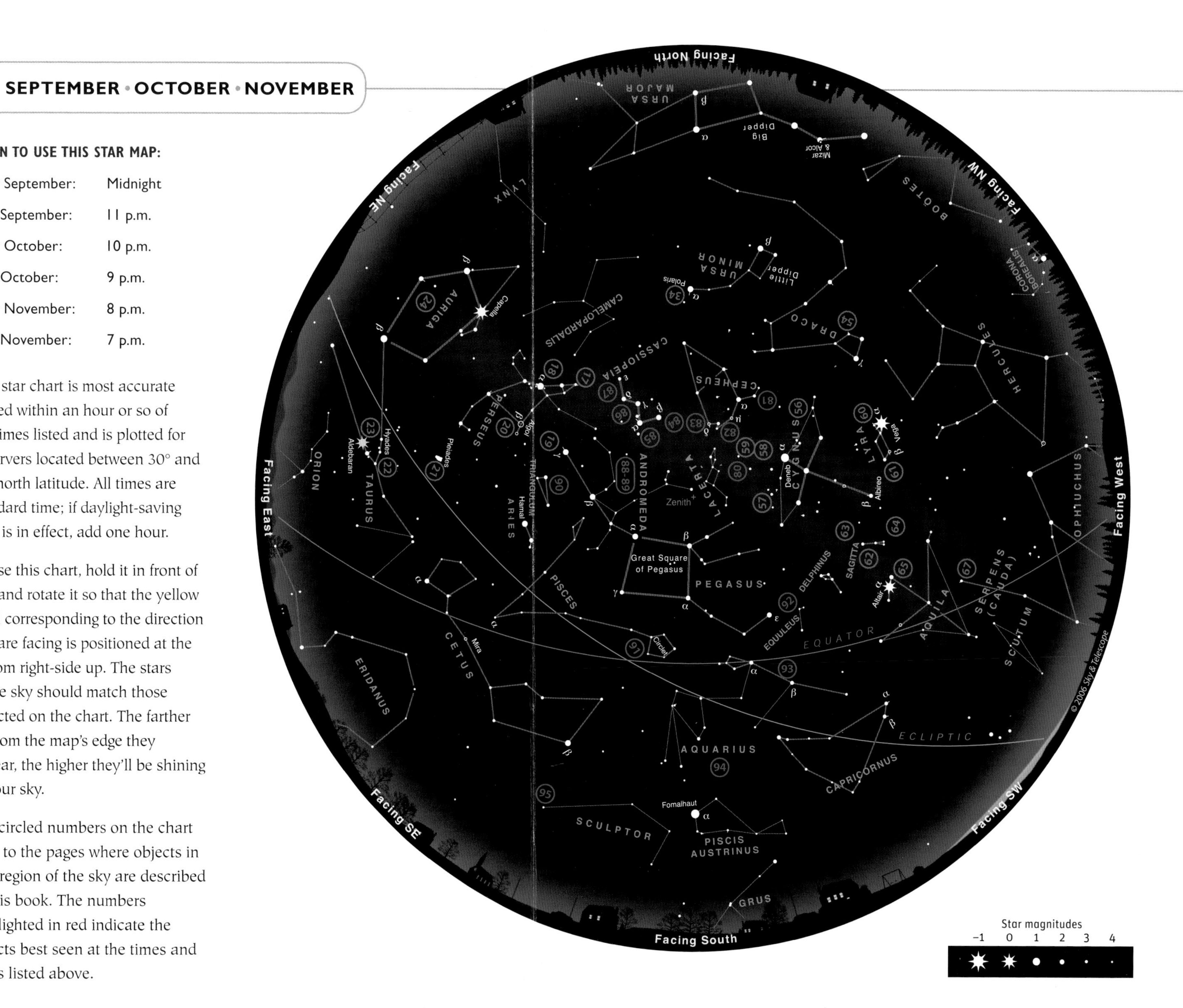

WHEN TO USE THIS STAR MAP:

Early June:	1 a.m.
Late June:	Midnight
Early July:	11 p.m.
Late July:	10 p.m.
Early August:	9 p.m.
Late August:	Dusk

This star chart is most accurate if used within an hour or so of the times listed and is plotted for observers located between 30° and 50° north latitude. All times are standard time; if daylight-saving time is in effect, add one hour.

To use this chart, hold it in front of you and rotate it so that the yellow label corresponding to the direction you are facing is positioned at the bottom, right-side up. The stars in the sky should match those depicted on the chart. Ignore all the parts of the map above horizons you're not facing.

The circled numbers on the chart refer to the pages where objects in that region of the sky are described in this book. The numbers highlighted in red indicate the objects best seen at the times and dates listed above.